Distance Learning and University Effectiveness: Changing Educational Paradigms for Online Learning

Caroline Howard
Techknowledge-E Systems, USA

Karen Schenk
K. D. Schenk and Associates Consulting, USA

Richard Discenza
University of Colorado at Colorado Springs, USA

 Information Science Publishing

Hershey • London • Melbourne • Singapore

KH

Acquisition Editor:	Mehdi Khosrow-Pour
Senior Managing Editor:	Jan Travers
Managing Editor:	Amanda Appicello
Development Editor:	Michele Rossi
Copy Editor:	Bernard J. Kieklak
Typesetter:	Jennifer Wetzel
Cover Design:	Michelle Waters
Printed at:	Integrated Book Technology

Published in the United States of America by
Information Science Publishing (an imprint of Idea Group Inc.)
701 E. Chocolate Avenue, Suite 200
Hershey PA 17033
Tel: 717-533-8845
Fax: 717-533-8661
E-mail: cust@idea-group.com
Web site: http://www.idea-group.com

and in the United Kingdom by
Information Science Publishing (an imprint of Idea Group Inc.)
3 Henrietta Street
Covent Garden
London WC2E 8LU
Tel: 44 20 7240 0856
Fax: 44 20 7379 3313
Web site: http://www.eurospan.co.uk

Library of Congress Cataloging-in-Publication Data
Distance learning and university effectiveness : changing educational paradigms for online learning / Caroline Howard, Karen Schenk, Richard Discenza, editor[s].
    p. cm.
Includes bibliographical references and index.
  ISBN 1-59140-178-X (hardcover) -- ISBN 1-59140-179-8 (ebook) -- ISBN 1-59140-221-2 (pbk.)
  1. Distance education--Computer-assisted instruction. 2. Education, Higher--Computer-assisted instruction. 3. Education, Higher--Effect of technological innovations on. I. Howard, Caroline, 1953- II. Schenk, Karen, 1955-   . III. Discenza, Richard.
  LC5803.C65D56 2003
  378.1'758--dc22
                    2003014956

British Cataloguing in Publication Data
A Cataloguing in Publication record for this book is available from the British Library.

All work contributed to this book is new, previously-unpublished material. The views expressed in this book are those of the authors, but not necessarily of the publisher.

10/25/04

Distance Learning and University Effectiveness: Changing Educational Paradigms for Online Learning

Table of Contents

Preface vi

Caroline Howard, Techknowledge-E Systems, USA
Karen Schenk, K. D. Schenk and Associates Consulting, USA
Richard Discenza, University of Colorado at Colorado Springs, USA

Section I: Strategies and Paradigms

Chapter I. How Distance Programs will Affect Students, Courses, Faculty and Institutional Futures 1

Murray Turoff, New Jersey Institute of Technology, USA
Richard Discenza, University of Colorado at Colorado Springs, USA
Caroline Howard, Techknowledge-E Systems, USA

Chapter II. Design Levels for Distance and Online Learning 21

Judith V. Boettcher, Designing for Learning, USA

Chapter III. E-Moderating in Higher Education 55

Gilly Salmon, Open University Business School, United Kingdom

Chapter IV. Community-Based Distributed Learning in a Globalized World 79

Elizabeth Wellburn, Royal Roads University, Canada
Gregory Claeys, Aboriginal Development Consultants, Canada

Section II: Course Development Instruction and Quality Issues

Chapter V. Online Course Design Principles 99
   Lance J. Richards, Texas A&M University, USA
   Kim E. Dooley, Texas A&M University, USA
   James R. Lindner, Texas A&M University, USA

Chapter VI. Theory and Practice for Distance Education: A Heuristic Model for the Virtual Classroom 119
   Charles E. Beck, University of Colorado at Colorado Springs, USA
   Gary R. Schornack, University of Colorado at Denver, USA

Chapter VII. Looking for Indicators of Success for Distance Education 144
   Wm. Benjamin Martz, Jr., University of Colorado at Colorado
      Springs, USA
   Venkateshwar K. Reddy, University of Colorado at Colorado
      Springs, USA
   Karen Sangermano, University of Colorado at Colorado Springs,
      USA

Chapter VIII. Online Assessment in Higher Education: Strategies to Systematically Evaluate Student Learning 163
   Elizabeth A. Buchanan, University of Wisconsin-Milwaukee, USA

Chapter IX. Assessing the Impact of Internet Testing: Lower Perceived Performance 177
   Wm. Benjamin Martz, Jr., University of Colorado at Colorado
      Springs, USA
   Morgan M. Shepherd, University of Colorado at Colorado Springs,
      USA

Chapter X. Modular Web-Based Teaching and Learning Environments as a Way to Improve E-Learning 190
   Oliver Kamin, University of Göttingen, Germany
   Svenja Hagenhoff, University of Göttingen, Germany

Chapter XI. The Effect of Culture on Email Use: Implications for Distance Learning 213
   Jonathan Frank, Suffolk University, Boston, USA
   Janet Toland, Victoria University of Wellington, New Zealand
   Karen Schenk, K. D. Schenk and Associates Consulting, USA

Section III: Building an Organization for Successful Distance Educations Programs

Chapter XII. A Strategy to Expand the University Education Paradigm 235
    Richard Ryan, University of Oklahoma, USA

Chapter XIII. Return on Investment for Distance Education Offerings: Developing a Cost Effective Model 253
    Evan T. Robinson, Shenandoah University School of Pharmacy, USA

Chapter XIV. Financing Expensive Technologies in an Era of Decreased Funding: Think Big ... Start Small ... and Build Fast 278
    Yair Levy, Florida International University, USA
    Michelle M. Ramim, MIS Consultant, USA

Chapter XV. Education Networks: Expected Market- and Cost-Oriented Benefits 302
    Svenja Hagenhoff, Georg-August-University Goettingen, Germany
    Michaela Knust, Georg-August-University Goettingen, Germany

Chapter XVI. Supplemental Web Sites: An Innovative Use of Information Technology for Instructional Delivery 322
    Malini Krishnamurthi, California State University, Fullerton, USA

About the Authors 340

Index 349

Preface

Universities are embracing distance education, yet most are not making the changes necessary to maximize the effectiveness and efficiency of online learning. Historically, universities have designed and built pedagogies, reward systems, organizational structures, procedures, and policies to facilitate face-to-face modes of education. University staff and faculty cling to deep-rooted paradigms, which may have worked well with traditional forms of education, but do not work well for technologically enhanced distance education. The majority of online courses utilize pedagogies developed for customary classroom learning environments. To compensate for the lack of face-to-face interaction, successful remote courses capitalize on innovative, technology-enabled methodologies to facilitate and enhance the student learning experience. Whereas conventional teaching methodologies do not routinely use the power of technology to assist scholarship, not using these methods may compromise remote course quality. By clinging to traditional pedagogies, universities often diminish the potential educational advantages brought by the technologies used for distance education.

Faculty evaluation and promotion processes reward behaviors desirable in conventional educational environments and fail to recognize and support behaviors needed in the online environment. University staff, administration, and faculty are organized to support traditional student populations. Many university policies and procedures rely on face-to-face interactions with students. Institutional systems are designed to meet the needs of face-to-face students. For example, it may be difficult for the distributed students to obtain help with technical problems due to information technology help desks which do not provide the services needed to meet the needs of distance students and faculty and the increased technical demands associated with online learning. Similarly, distance students may not have access to academic advising and other student support services.

Distance learning is a good example of a student-centered program, because it accommodates the lives and lifestyles of students who have jobs and families, but wish pursue higher education. Distance education allows students a unique opportunity to pursue a college education, live at home, maintain a career and enjoy the convenience of anywhere, anytime learning.

Accomplishing high quality distance education will mean that university administrations, faculty, and other constituents will have to make changes. Making the changes necessary to accommodate effective distance education will be difficult. Carol Twigg, a leader in the distance education field observes:

"Leaders of the old paradigm have a tremendous amount of time and energy invested in using the old rules. Consequently, they are often resistant to change and less likely to look for creative, innovative approaches to new opportunities. In much the same way that Thomas Kuhn (who first called our attention to the idea of paradigm shifts) observed scientists trying to 'save the theory', so too do defenders of the old paradigm focus their efforts on old solutions to new problems."

To succeed in distance education, faculty members must be willing to change their teaching methods and their reward expectations. Universities will need to transform their structures, rewards, and policies to accommodate the needs of distance education programs. In summary, the influx of distance education into the system is forcing universities to rethink their foundations and shift their paradigms.

This book discusses the many challenges presented in the distance education environment, explains needed paradigm changes, portrays innovative approaches for meeting these challenges, and describes the unique opportunities brought by distance education technologies. It shares some of the experiences that universities have undergone or are currently undergoing to take advantage of technology-enabled distance learning.

Organization of the Book

The book is organized into three sections. The first section presents strategies and paradigms for creating successful distance education programs. In this section, the authors use their vast experiences with diverse distance educational venues to formulate approaches for facilitating optimal learning experiences. These distance education pioneers share valuable insights into how to design methodologies and structures to enhance remote learning. The authors' diverse experiences with distance learning situations enhance the broad perspective that allows a multitude of design strategies for success.

Section I focuses on strategies and paradigms for distance education. In Chapter 1, the authors describe how, designed properly, distance education classes can be as effective, or even more effective, than face-to-face courses. The authors discuss how the tools and technologies used for distance education courses facilitate learning opportunities not possible in the face-to-face classroom. They also describe the many changes required by students, faculty, and the university itself to meet these new challenges. Changes in reward systems, skill sets, and faculty commitment to excellence in teaching necessary for successful paradigm shifts are discussed.

Chapter 2 presents a framework with six levels of design for online and distance programs. This multi-leveled design process is grounded in learning and change theory, as well as instructional design. This chapter provides a set of principles for designing effective and efficient online and distance learning programs.

Chapter 3 describes a low cost approach, *e-moderating*, for developing lectures for new online roles. This chapter considers and explores the knowledge and skills that the best e-moderators have and how they can be recruited, trained and developed.

In Chapter 4, the authors illustrate how respectful partnerships can be developed where communities control how their knowledge is used and retained while tapping into the potential of new technologies and maintaining an appropriate level of quality. They demonstrate the potential of this philosophy by examining two pilot programs for indigenous learners.

Section II focuses on instructional, course development and quality issues. Moving from traditional classroom to online and distance learning programs takes innovation and careful planning. The authors in this section suggest indicators and strategies for maintaining quality throughout the design and implementation process.

Chapter 5 discusses the fundamental changes necessary for instructors to make distance learning more effective and appropriate for a growing audience. The authors present the areas of competence important for distance teaching, such as course planning and organization, verbal and nonverbal presentation skills, and questioning strategies. They also provide practical guidelines and examples for designing online courses.

Chapter 6 adapts systems theory to distance education. Using an educational process model, the authors examine inputs, process and outputs that lead to a list of best practices for tackling the virtual classroom. Internal and external assessment guidelines help direct successful outputs in the distance learning process.

Chapter 7 identifies key components of distance education satisfaction based on previous research by comparing traditional and distance education environments. The authors administer a questionnaire based on these components to 341 MBA distance students and factor the results into five constructs that correlate well with the satisfaction ratings of the subjects.

Chapter 8 acknowledges the challenges surrounding assessment techniques in online education at the higher education level. It asks specifically, "How do we know our online students are learning?" To answer this question with confidence, various strategies ranging from participation techniques to online group work, peer and self-assessment, as well as journals and portfolios, are described.

In Chapter 9, the authors describe a comparison between two sections of a graduate programming class, where one is an on-campus class and the other a distance class. Differences in perceived test performance were found between sections. Possible causes, implications, and suggestions for future research are explored.

In Chapter 10, the authors contend that conventional electronic learning materials are only suitable for a small target audience and cover only a few learning styles. They provide examples of the development of high quality and economically sensible distance learning materials. After discussion and evaluation of conventional materials, the authors guide the readers through the development of online study materials using modular construction. Necessary consideration of multiple interests and learning styles are discussed.

Chapter 11 looks at the effect of cultural differences on e-mail usage. Differences in use of technology-enabled communication can have large implications in the distance learning process. The authors examine how students from Collectivist versus Individualistic cultures utilize e-mail communication in distance education courses and the challenges these differences present.

Section III examines a variety of issues and strategies for building successful distance learning programs and the organizations that support them. The authors in this section discuss changes from conventional thought concerning funding, building and delivery for online programs. Starting small, building fast, innovative technology use, cost-oriented benefits, and return on investment are all issues discussed.

In Chapter 12, the author suggests a variation from the traditional approach for packaging and delivering Internet education. One strategy is to look beyond the "class" delivery approach. The premise for this strategy is the belief that the greatest strength of the Internet for education may lie in delivery of class "components," not classes, themselves. Online components can be used, not to replace, but to supplement and add value to the traditional class experience. This strategy proposes that the universities provide, sponsor, administer and maintain an automated online portal to post and sell faculty created material.

Chapter 13 identifies ways in which institutions can maximize their return on investment for distance education offerings through the appropriate and timely re-purposing of the online content for different markets. A transformative income generation (TIG) model is suggested as a means of generating additional revenue by leveraging existing assets. Targeted application of this model is used to maximize return on investment for online distance education development.

Chapter 14 examines two common approaches that higher education institutions pursue when implementing online learning programs and provides the rationale for their success or failure. The authors define, propose, and categorize a set of eight key elements of a successful online learning program implementation in an era of decreased funding. The chapter also contains a case study involving the development of a successful self-funding online learning program in the college of business administration at a southeastern state university.

Chapter 15 describes an inter-university education network and the expected effects of co-operational activities. This chapter focuses on both cost-oriented and market-oriented benefits of the education network WINFO*Line*. Additional cost and market advantages of WINFO*Line* are discussed also. The authors initiate further discussions by looking at some problems and questions about managing open education networks in a co-operational environment.

Chapter 16 presents the results from an empirical study that show students perceive a face-to-face course supported by a web site to be useful in enhancing their academic performance. Almost all the students made use of the classroom lectures and web site resources, without feeling the need to stay away from lectures. Since learning can now be done synchronously or asynchronously, individually or collaboratively, at any time or place through the use of technology, the task of incorporating technology judiciously into the learning environment becomes a new challenge. A plethora of opportunities exist and those choices come with varying degrees of challenge and success.

Acknowledgments

The editors would like to acknowledge the many people who made this project possible. We appreciate the help and cooperation of the chapter authors included in this volume. We appreciate the quality of their work and it has been a pleasure working with them. Many served as referees for chapters submitted by other authors and took the time to provide comprehensive reviews and constructive comments. These were invaluable during the final selection and revision process. We are enormously grateful to Evan Robinson and Elizabeth Buchanan for taking on the extra reviews needed to get this work accomplished. We especially want to thank those authors whose chapters have contributed to the quality of both of our distance education books: Kim Dooley, M. Ben Martz, Morgan Shepherd, Elizabeth Buchanan, and Richard Ryan.

We want to acknowledge the assistance of the talented staff at Idea Group Publishing, also. We extend our thanks and appreciation to all of the editors and staff at Idea Group Publishing, without whom the writing of this book would not have been possible. The editorial staff's support and quick responses to our inquiries helped facilitate the process and keep the project on schedule. We especially appreciate Michele Rossi's guidance and help in managing the project from start to finish. Jan Travers' editorial assistance was key in producing a high quality product. We benefited from Mehdi Khosrow-Pour's guidance in the early formulation of the book. Carrie Skovrinskie's correspondence and efforts kept the marketing plan moving forward.

The preparation of a book of this kind is dependent on excellent colleagues and the three of us have been fortunate in this regard. Through inspiration and encouragement, it has been a team effort from start to finish.

Last, but not least, I, Caroline, want to thank my husband and my children for their love, support, and patience during this project. I thank Devon Pasquariello for excellent editorial suggestions. I also am grateful to Karen and Dick for being the best possible partners on both books.

From Karen to Katie and Robert, whose encouragement and love has allowed me to complete another project with interesting and inspiring colleagues — thank you, Dick and Carol, also.

From Dick, as always, I am especially grateful for the support of my co-editors, friends, associates, and family. The time and energy spent on editing and writing this, plus other articles and books has been greatly facilitated by their encouragement, patience and understanding.

Caroline Howard
Karen Schenk
Richard Discenza

Section I

Strategies and Paradigms

Chapter I

How Distance Programs will Affect Students, Courses, Faculty and Institutional Futures[1]

Murray Turoff
New Jersey Institute of Technology, USA

Richard Discenza
University of Colorado at Colorado Springs, USA

Caroline Howard
Techknowledge-E Systems, USA

Abstract

Designed properly, distance education classes can be at least as effective and, in some ways, even more effective than face-to-face courses. The tools and technologies used for distance education courses facilitate learning opportunities not possible in the face-to-face classroom. Distance programs are accelerating changes that are challenging students, faculty, and the university, itself. Currently, most faculty are rewarded for the quality of instruction, as well as their external funding and their research. Often, university administrators focus more attention on the efficiency of teaching than on its effectiveness. In the future, as the quality of distance learning increases, the primary factor for success will be the faculty's commitment to excellence in teaching. Many institutions will be forced to reevaluate the quality of teaching as the institution becomes more visible to the public, to legislators who support higher education, and to prospective students.

Introduction

People usually assume that students in distance education programs are at a disadvantage. On the contrary, it is probably not the distance student who is disadvantaged, but rather many face-to-face students. Learning is enhanced by the physical and social technologies typically used in distance education. Students in distance programs have access to tools that allow them to repeat lectures and interact with their fellow students and faculty. Contrast these students with a student sitting in a 500 student lecture. Which student is most at a distance?

In the early 1980s, a research group introduced a computer-mediated system to a regular face-to-face class. The group felt that there was enormous potential for this technology to enhance learning. The system was introduced to students in a number of Computer Science and Information System courses. Due to the amount of material covered in lectures, there was not much time for dialogue and only a few students participated when there was a class discussion. The instructors introduced asynchronous group communication technologies to communicate discussion questions and assigned grade point credits for student participation. One hundred percent of the students participated in these discussions outside of regular classroom hours. The extent and depth of the discussions changed the nature of the classes. Most importantly, student contributions were comprehensive, with more well-thought-out comments, because students had the time to reflect on the ongoing discussion before participating. Also very significant was that students, for whom English was a second language, became equal participants. They could reread the online discussion as many times as needed before replying. The computer-based activity monitoring and transcripts, electronic recordings of the discussions, showed that foreign students spent two to three times more in a reading mode and reread many discussions, far more than the American students.

This ability to monitor activities and review the electronic transcripts gives the instructor insights into how students are learning. By reviewing the transcripts of the online discussions, it becomes obvious what and how students are learning.

For courses with high pragmatic content, such as upper level and graduate courses in topics like the design and management of computer applications, students are required to utilize problem-solving approaches to evaluate the trade-offs between conflicting objectives. In a traditional classroom environment, especially in large classes, it is very difficult to detect whether students are accurately incorporating the problem-solving mental models that the instructor is attempting to convey. Reviewing the transcripts of class discussions can provide insight into the approaches students are taking to master the material.

Unfortunately, in the early 1980s no one wanted to hear about a revolution in normal classroom teaching or was willing to expend the effort to dramatically improve classroom education. It was only those interested in distance education who were interested in learning about the educational potential of the technology. As a result, in the mid 1980s the researchers at New Jersey's Institute of Technology (NJIT) obtained research funding to investigate distance education applications of Computer Mediated Communications (CMC). Since NJIT, at the time, had no distance program they created distance sections of regular courses that were used with regular on-campus students taking most other courses face-to-face.

This effort (Hiltz, 1994) utilized quasi-experimental studies that compared a population of students (only familiar with face-to-face classroom education) to a population of students taking the same courses in pure face-to-face sections with pure distance sections using only CMC technology. The students in the matched sections had the same material, the same assignments, the same exams and the same instructor. They found no significant difference in the amount of learning or the rate of student satisfaction. This finding is much more significant than a determination based on a study that included a population of distance learners already familiar with traditional correspondence classes. Two critical underlying variables driving the success of this approach were identified by Starr Roxanne Hiltz (Hiltz, 1994). First, the role the instructor needed to take was different from the traditional classroom role. The instructor acted more as an active and dedicated facilitator rather than traditional teacher, as well as a consulting expert on the content of the course. Second, collaborative learning and student teamwork were the educational methodology (Hiltz, 1994) which was shown in later studies to be a key factor in making distance courses as good or better than face-to-face courses (Hiltz and Wellman, 1997).

These results indicate that distance courses can be as effective as face-to-face courses when using any of the traditional measures, such as exams and grades. However, these traditional measures may be inadequate to measure of many of the benefits observed in classes that utilize computer-mediated communications technology for a number of reasons:

- Due to social pressures, students tend to be more concerned with how other students view their work quality than how the professor views it. They are significantly more motivated to participate in a meaningful way when their fellow students can view their contributions.

- When equality of communications is encouraged, students cannot get away with being passive or lazy. The transcript or electronic recording of the discussions shows who is and is not participating. It is visible to both the instructor and other students that someone is being lazy. (In fact, students seem to be more concerned with what the other students will think of their performance than what the professor will think.)

- The scope of what the outstanding students learn becomes even more noticeable.

- The performance of students at the lower end of the distribution is improved. The communications systems permit them to catch up, because they are able to obtain a better understanding of the material with which they are most uncomfortable or have the least background knowledge.

- The instructor can become more aware of his/her successes or failures with individual students because of the reflective nature of the student contributions to the discussion.

While these conclusions need confirmation through long-term longitudinal studies of student performances, the marketplace is also providing confirmation of the beliefs held by many experienced in teaching these classes. We are seeing that collaboratively oriented programs offer a solution to the problems, which are inherent in traditional correspondence courses.

Students benefit from the ability to electronically store lectures alone or in chunks integrated into other material on the Web. Electronic storage of lectures gives all students the power to choose freely whether they want to attend a face-to-face class or take the same course remotely. Traditional face-to-face students can later hear a lecture missed due to illness or travel. Students with English as a second language can listen to a lecture multiple times. Face-to-face students who have to travel or fall sick can use the same tapes to catch up and/or review material prior to exams.

In our view a student in a face-to-face class that is not augmented by a collaborative learning approach and by asynchronous group communications technology is not getting as good of an education as the distance student who has those benefits. It is the face-to-face student who may be suffering from the segregation of the college system into separate face-to-face and distance courses. These observations about the past and the present lead to some speculations about the future.

The Changing Nature of Academic Courses

A vital role of college education is to convey the mental model and pragmatics of a subject matter. To ensure learning of complex subject matter, an instructor needs to communicate his/her mental model and accompanying problem-solving approaches. The model, or structure and dimensions to understand and organize the material, along with the pragmatics of the subject matter become the student's starting point. These allow him/her to build the details, acquire new knowledge, and apply information in new situations. The more pragmatic the course content is, the more important it becomes for the instructor to convey his or her mental models and assess student assimilation.

Communication among students and with the instructor is particularly valuable in courses with high pragmatic content. By discussing and comparing their interpretations, student can reinforce their understanding and reduce their conceptual errors. Mixing distance and face-to-face students in the same discussion space results in having students with a great deal of work experience virtually mingle with those who have had none. Often undergraduate students who take distance courses are working, and most of those in the face-to-face courses are not working. Mixing both a distance and a face-to-face class leads to a better balance of backgrounds. Students, who have work experience, may have had an example from their life that illustrates a theoretical concept presented in the class. When a student shares this experience, it reinforces the importance of the concept and encourages other students to pay attention to the presentation.

Creative, interactive software programs accompanied with background tutoring can effectively teach students to master the skills currently taught in many undergraduate courses. When these courses are automated, the costs incurred are far below typical college tuition. In the future, colleges and universities will not be able to continue to charge current tuition costs for introductory courses that are largely skill oriented. For example, there are many stand-alone and Web-based software programs that offer introductory programming courses, as well as skills in many other areas. These courses are comparable to college courses and some are even based upon a textbook used on some college campuses. They are available for a few hundred dollars. The major difference is that they do not carry college credits.

At the other extreme we have the usual "skimming of the cream" that occurs with every change in technology. Today, private firms are willing to invest a millions dollars in single multimedia, largely automated courses to sell to industry that can afford to pay thousands of dollars of tuition for each student. When all the downsizing of outdated professionals was occurring, one of the first things to go was the elaborate investment these companies had made in internal course offerings and video classrooms and networks. Some companies had created their own internal college and claimed their employees did not need outside education. This type of thinking, like the concept of "just-in-time learning," which may be a euphemism for "teach them only what they need to know in order to do their current job," does not prioritize student growth and learning. Unfortunately, some institutions of higher education are no longer certain about whether their client is the student or industry.   Until an enlightened consumer marketplace emerges, many transient but inferior offerings will be available on the market.

With the recent recession we find some institutions, like Duke, have cancelled their online executive M.B.A., for which they were charging twice the tuition of the on-campus offering. The recession has had the obvious results of discouraging and terminating a number of private sector efforts to enter the distance education market. The cream has largely turned sour.

Now that we have a national and international competitive market in courses, those colleges that accept skill knowledge from unaccredited sources, such as training courses and work experience, will obtain a market edge over those that do not. In fact, the student population will begin to expect institutions of higher learning to accept courses from any accredited institution.

In addition, institutions with clearly stated credit transfer policies will also obtain a marketplace edge. Individual courses, as well as total programs, will be the basic units in a national and international marketplace for higher education. There are no longer geographical monopolies on higher education. Only consortiums based upon real cooperation among the participating institutions will succeed. Many current attempts to market only the separate offerings of the participating institutions, or to impose added layers of administration between the courses and the students, are doomed to a marketplace failure.

The single noncommercial web site that focuses on distance education utilizing group communications is that of the Society for Asynchronous Learning Networks (http://www.aln.org). There are numerous group conferences for educators and administrators, as well as a newsletter and journal.

Role of the Faculty

In any meaningful educational program, the major responsibilities for a given course should be the responsibility of a faculty member, including:

- Course design
- Choice and creation of course materials and assignments

- Approval and mentoring of instructors, adjuncts, and Ph.D. students who will also teach the course
- Performance monitoring of other instructors, adjuncts, and Ph.D. students
- Course and material updates

The technology allows senior professors or department chairs to effectively evaluate and mentor all instructors of particular courses, whether they are teaching traditional classroom courses or distance courses. The ability to review whole class discussions after the class is over gives senior faculty the ability to evaluate distance instructors hired to teach previously developed courses, as well as to review on-site instructors and junior faculty. Thus, they can improve and extend their mentorship and apprenticing relationships.

While educational institutions are rapidly developing programs for large populations of distance students, it does not appear that universities are creating tenured faculty lines that can be occupied by remote faculty. When additional faculty are needed to teach distance courses, instructors, rather than tenure track faculty, are often sought. Since the success of distance courses is largely dependent upon the capabilities of the instructor responsible for the particular course, the value of instructors able to master teaching at a distance will rise within institutions of higher education.

The technology we are using for distance education can allow faculty members to live anywhere they want to. Unique benefits will be available to outstanding teaching faculty. For example, one of the best full-time instructors for NJIT, which is located in beautiful downtown Newark, is a mother with two small children who never has to be on campus. She is teaching other instructors how to teach remotely. Similarly, a University of Colorado accounting professor, on sabbatical in Thailand, is able to teach a course in the Distance M.B.A. program.

There have been a few master programs where some or all of the instructors are located anywhere in the world. It is technically feasible for those wanting to escape winter cold to teach in places such as Hawaii that we could only dream about. The technology makes it feasible, but various administrative policies, unions, insurance companies, benefit programs, etc., have not yet caught up to the technology. There is increasing emphasis by accrediting agencies on treating remote instructors the same as faculty are treated. This is likely to bringing about a greater degree of equality between instructors and tenured track faculty. The outcome is uncertain, but it may mean that the costs for remote and traditional classes will equalize so that the profit margin in online classes will not be quite so high.

Role of the Technology

Some functions of technology that can facilitate this function are:

Asynchronous discussions: In the online environment, students can take as much time as they need to reflect on a discussion and polish their comments. This improves the quality of the discussion and changes the psychology and the sociology of communications. Students can address topics in the sequence they chose rather than in a predefined order. This leads to the development of different problem-solving strategies among the individual members of the class.

Instructor control of online conference and roles: With online course conferences (many per course), instructors control the membership of each, assign roles and enable other instructors to monitor conferences for joint teaching exercises involving more than one course. Groups within courses are able to set up private online conferences for team and collaborative work group assignments. Joint editing of items facilitates team work.

Question and answer communication protocol: Instructors are able to ask questions during discussions. They can control who views the answer and prevent other students from seeing the answer of the others or engaging in the resulting discussion until they have entered their answer. In studies of Group Decision Support Systems, it has been shown that asynchronous groups in an online Delphi mode generate many more ideas than unstructured discussions or face-to-face groups of similar size (Hee-Kyung et al., 2003). This area has proven to be a valuable tool in forcing equal participation. Use of question-and-answer communication protocol can be used to force each student to independently think through their answer without being influenced by the other students.

Anonymity and pen name signatures: When students with work experience are part of a discussion, they can use their real life experiences to illustrate the concepts the professor is presenting. Such comments from fellow students, rather than the professor, often make the instructor's message more meaningful to the students. A student confirming the theory presented by a faculty member through real life examples is more effective in making a point than "dry" data from an instructional article. Furthermore, students can talk about disasters in their companies with respect to decisions in any area and provide detail, including costs, when they are not identified and the anonymity of the company they work for is preserved   Also, the use of pen names allow individuals to develop alternative personas without divulging their real identity and is extremely useful in courses that wish to employ role playing as a collaborative learning method.

Membership status lists: The monitoring of activities, such as students' reading and responding to communications, allows the professor to know what each individual has read and how up-to-date each student is in the discussion.

This allows the instructor to detect when a student is falling behind. Student collaborative teams can make sure that everyone in the team is up to date. Furthermore, students can easily compare their frequency of contributions relative to other students in the course.

Voting: Instant access to group and individual opinions on resolutions and issues are enabled by voting capabilities. This is useful for promoting discussion and the voting process is continuous so that changes of views can be tracked by everyone. Voting is not used to make decisions. Rather, its function is to explore and discover what are the current agreements and disagreements or uncertainties (polarized vs. flat voting distributions) so that the class can focus the continuing discussion on the latter. Students may change their votes at anytime during the discussion.

Special purpose scaling methods: These useful methods show true group agreements and minimize ambiguity. Currently we have a system which allows each student at the end of the course to contribute a statement of what they think is the most important thing they have learned in the course and then to have everyone vote by rank ordering all the items on the list. The results are reported using Thurstone's scaling, which translates the rank order by all the individuals to a single group interval scale. In this interval scale if 50% prefer A to B and 50% prefer B to A, the two items will be at the same point on the scale. It has been surprising what some of the results have been in some courses. For example, in a management of Information Systems course the concept of "runaway" software projects was felt to be twice as important as any other topic. The professor was quite surprised by this result until he began to realize that the students were using this concept of a mental model in which to integrate many of the other things they had learned.

Information overload: This occurs when enthusiastic discussions by students that are meant to augment the quality of the learning process augment only the quantity of the number of comments, instead, leading to the problem of "information overload." Currently this phenomenon limits the size of the group that can be in a single CMC class. Online discussions allow individuals to enter comments whenever it is convenient for them, without waiting for someone else to finish the point they were trying to make. This makes it physically possible and also very likely that a great deal more discussion will take place and much more information will be exchanged among the group than if only one person can speak at a time, as in the face-to-face classroom environment. Anything that reduces the temptation of some students to "contribute" comments or messages that have nothing to do with the meaningful discussions underway will increase the productivity of the discussion without information overload setting in. Among such functional tools the computer can provide are:

Class gradebooks: This eliminates a tremendous amount of electronic mail traffic that would become very difficult for an individual instructor to manage with a large class.

Selection lists: The instructor can set up lists of unique choices so that each student may choose only one item and others can see who has chosen what. This is very efficient for conveying individualized assignments and reduces a large portion of communications.

Factor lists: Members of a class or group can add ideas, dimensions, goals, tasks, factors, criteria, and other items to a single, shared list which may then be discussed and modified based upon that discussion and later voted upon.

Notifications: Short alerts notify individuals when things occur that they need to know about. For instance, students can be notified that a new set of grades or vote distribution has been posted, eliminating the need for individuals to check for these postings. People can attach notifications to conference comments from a select list that provides alternatives like: *I agree, I disagree, I applaud, Boo!* Such appendages reduce significantly the need to provide paralinguistic cues of reinforcement as additional separate comments.

Calendars, agendas or schedules: Students have access to a space to track the individual and collaborative assignments and their due dates. These are listed in an organized manner that links detailed explanations for each assignment, as well as questions and answers related to the assignments.

The State of the Technology

The technology available today includes at least 250 versions of group communication software. However, some of them may not survive the recession. There are a growing number of software packages for course management. The online learning product landscape is changing at a rapid pace as companies are acquiring their competitors to expand functionality. A recent article gives an excellent summary of the popular platforms and the evolving nature of eLearning (Gray, 2002).

There are only a few of these that have wide usage and they are beginning to raise their prices to capitalize on their popularity. Most of these packages charge a fee per user, which is not the desirable fee structure for the customer. Many of the older conferences systems charge on a per server basis and it does not matter how many students one has. It is far cheaper to spend more on the hardware and a get a more powerful server. Also, the course management systems do not provide many of the useful software features one would like to have for group communications. Given the way prices are going, it might be better to pay some of the undergraduate students to educate some of the faculty on how to create their own web sites and have their own pages for their courses that they update and maintain directly. This also has desirable long-term consequences in raising the ability of the faculty in this area. Once you have committed all your content to one vendor's system, you are a captured customer and will have to pay whatever they want to charge. Right now, software development is undergoing rapid evolution and no customer should put themselves in the box of only being able to use one vendor. If it is clear you are using a number of vendors, you may even be able to get some breaks on pricing and will certainly get the top level of service when each of them knows there is an alternative service readily available to the customer. In the coming decade, one can expect major upgrades of these software systems every few years and the best one today may not be the best one tomorrow.

Course Development and Delivery Technology

Unfortunately, many faculty do not know how to use the technology to design a successful course. As the historical record shows, it is a mistake when transferring an application to computers to just copy the way it used to be done onto the computer. Utilizing the methodology of collaborative learning is the key to designing courses using group communications technology. Simple systems, which attempt to impose a discussion thread on top of what is electronic mail technology, allow the student or the teacher only to view one comment at a time. This approach does not allow an individual to grasp the totality of any complex discussion. Only by placing the complete discussion thread in a single scrolling page can a person review and understand a long discussion. They can browse the discussion and cognitively comprehend it without having to perform extra operations and loose their cognitive focus. Users of such simple systems cannot generate a large complex discussion and have no way of realizing that complex discussion is even possible.

When online discussions are successful, they can easily go from enthusiastic wonderful discussions to information overload. To maximize the power of the technology to facilitate collaborative learning, critical development directions for the future should include:

- Tailorability of communication structures by instructor
- Tailorability of communication protocols by instructor
- Anonymity and pen name provisions
- Delphi method tools and the availability of scaling and social judgment (voting methods)
- Tools for collaborative model building
- Powerful information retrieval capabilities
- Tailorability by instructor of application-oriented icons and graphical components
- Tools for the analysis of alternative diagrams

Instructors also need to allow students to extend the discourse structure and to vote on the significance of incidents of relationships among factors in the problem domain by using Group Decision Support processes. The system should allow students to not only develop their own conceptual maps for understanding a problem, but also to detect disagreements about elements of the conceptual map and the meanings of terms. This is valuable preparation for problem solving in their professional life, a process that requires removing inherent ambiguities and individual meanings in the language used to communicate about a problem with others from diverse backgrounds.

Routines should be included that are based upon both scaling and social judgment theories which improve the ability of larger groups to quickly reach mutual understanding. Currently, few tools exist in current systems that support the use of collaborative model building, gaming, and Delphi exercises. The current generation of software does not often include the functions of anonymity and pen names.

Course instructors need to have complete control over course communication structures and processes and should be able to use their recently acquired knowledge for future offerings of the course. Currently, systems lack the needed integration of functions to easily evolve the changes in both the relationships and the content in a given field. A long-term advantage of teaching in the collaborative electronic environment is that the students create useful material for future offerings and can aid the instructor in monitoring the new professional literature.

Future technology will allow faculty to organize their material across a whole set of courses into a collaborative knowledge base available to the faculty teaching those courses. This would allow students and faculty would be able to create trails for different objectives and weave the material in that knowledge base to suit a group of students or a set of learning objectives. Individual learning teams would be able to progress through a degree program's knowledge base at the rate best for them, rather than setting the same timeframe for all learning teams or faculty teams. Faculty, individuals or teams would take responsibility for a specific domain with in the web of knowledge representing a degree program.

Collaborative technologies are changing the concept of what constitutes a course. Program material could be an integrated knowledge web based largely on semantic hypertext structures. Over time, the domain experts, the faculty, would continue to develop and evolve their parts of the web and wait for learning groups, composed of any mix of distance and regular students sharing the same learning objectives and needs.

Current vendor systems focus on the mass market and concentrate on tools to standardize and present course content. Group communication tools are usually just disguised message servers that offer only a discussion thread capability and little more, certainly not the complex capabilities discussed above. Vendors have not yet recognized the primary importance of group communications and how faculty members can guide and facilitate the process and be available for consultation as needed. Based upon the conceptual knowledge maps they design, faculty should be encouraged to develop content structures that are character-istic of their subject matter. In the end, faculty should have the ability to insert group communication activities anywhere in their professional knowledge base (e.g., question/answers, discussion threads, lists, voting, etc.).

Educational Consumerism

Most of today's distance education course offerings give the illusion of differ-ence by placing materials on the Web, instead of providing them through the mail. These online course offerings still use the correspondence course model. These offerings typically include e-mail systems for one-on-one communication be-tween an instructor and an individual student, but do not include effective group communications featuring course content and delivery methodologies reworked for a distance-learning environment. E-mail is better than nothing, but no one would claim that e-mail is preferable to a face-to-face college course. The typical consumer of distance education does not understand the difference between courses with only simple e-mail systems and courses that have introduced sophisticated group communication processes.

Students in the United States pay the equivalent of the price of a used or new car every year to attend college. A student and his family make a major financial investment for a college education. Reports evaluating new and used automobile models and car-buying comparison web sites enable people to obtain detailed information on any model of car for little or no money. We predict the emergence of a successful "consumer report" organization for distance learning, similar to the college guides appearing each year, that will provide details about individual courses and instructors in different programs, unfiltered and direct from other students. (Some guides have already appeared, but some of them make money by charging the schools to list their programs.) There is, as of yet, nothing comparable to *Consumer Reports* or even the yearly *U.S. News & World Reports* independent rating of universities. Examples of sources now available are *Bears' Guide to the Best Education Degrees by Distance Learning* by John Bear, Mariah Bear, Tom Head, Thomas Nixon Ten Speed Press (ISBN: 1580083331; September 2001) or *Guide to Distance Learning Programs 2003* (*Peterson's Guide to Distance Learning Programs*, 2003) Petersons Guides (ISBN: 0768908191Bk&Cd-Rom edition; October 2002).

Ultimately, these sources will support more intelligent consumerism about college education. Since distance education eliminates a college's geographical monopoly, colleges will be forced to be much more sensitive to consumer pressures than they have been in the past. Since no educational institution or organization has had the foresight, so far, to do this, there is now a commercial Web firm that sells books, student services and other products and has committed to putting up a recommendation system to evaluate any distance course anywhere and have the results made Web accessible.

Today, university faculty are not rewarded as much on the quality of instruction as on their research and external funding. To obtain tenure and promotions, their instruction quality merely has to be acceptable and new faculty cannot afford to prioritize exceptional, innovative teaching. The problem is often with university administrations that focus their attention more on efficiency rather than effectiveness when it comes to teaching.

Instructors and Rewards

Administrations place most of their priority on competitive research and sponsored funds. Many face a rude awakening, as they are less in touch with the fact that the educational process is undergoing an unanticipated, unexamined, fundamental change. Some of these institutions will not realize until it is too late to change entrenched attitudes and bureaucratic processes fast enough to compete in the new competitive environment. During the next decade, many institutions may fail as a result of intense international and national competition.

In the future, the underlying factor of success will be the faculty's commitment to excellence in teaching and the quality and talent of the instructors. Many institutions will be forced to reevaluate their faculty incentives and the relative importance of teaching and research. Marketplace mechanisms will also force this reevaluation, as the relative quality of teaching for each institution becomes more visible to the public and prospective students. These pressures will force the threshold for acceptable teaching quality definitely to rise.

As the quality of distance learning increases, the view of distance education courses as inferior to traditional classes will disappear. It will become the talent of the instructor and his or her facility that will determine success. Additional organizational layers of intermediaries will doom a program to failure. For example, the majority of students taking distance courses in the future will be regular students who schedule a mix of distance and face-to-face courses to accommodate their schedule, family commitments, work commitments and their desire to complete their education in a timely manner. Their course ratings do not distinguish distance courses and face-to-face courses, but distance courses are even rated higher in many ways.

Alternative Versions of the Future

Students of the future will have many choices, the spectrum of which can be illustrated by examples of two students at extremes:

The Positive Future for the Student

After careful consideration of my options, I decided to turn down a scholarship from an Ivy League university to go to eU (Electronic University). I wanted to continue learning my family's business and did not want to move away from my fiancée. I was able to easily make up for the cost of the eU tuition, which was about 25% of the traditional school's, even with the recent 15% tuition reduction which that school just announced. I was convinced when I discovered that eU was rated as highly as the Ivy League school for the quality of its courses. The response of more than three million students in the "Learning Consumer Database" made the results for most of my courses statistically significant to the .05 significance level. Also, I found the comments of the professors about their courses much more extensive in the eU course ratings. Hardly any of the other school's professors responded.

All of that encouraged me to check the resumes of most of the eU professors. The results were quite surprising. Most of them are retired from other universities where they had tenure before they came to eU. They are all paid the same salary of $150,000 and work out of their homes all over the world. One of them wrote that 95% of his time and effort is devoted to instructional activities because eU has eliminated all committees except one to determine the departmental curriculum.

The eU classes usually range from 20 to 50 students with a great deal of emphasis on class discussion and collaborative work. The profile of students shows that over 70% are working in professional areas related to their degree programs. As a result, class discussions are high caliber. I got to actually eavesdrop on some ongoing courses once I had submitted an application.

I can take my exams at the local community college, which has a franchise from eU, and I can also use their sophisticated multimedia computers. eU accepts courses from any college and university that has accreditation for the same degree, so I can use another distance program when the course I need is closed at eU or not offered that semester, without pre-approval. With the three-semesters-per-year program, I can move a lot faster than in most two-semester programs.

I am concerned, however, with getting in, as I am fresh out of high school and they take very few students like me. Their rejection rate is much higher than most Ivy League schools'. I have tried to convince them that my four years of part time work in the family business should be counted. I hope that helps.

The Negative Future for the Student

I have decided to apply to eU.com rather than Harvard. I really must spend the time learning the family business and my fiancée has told me in no uncertain terms that long separations are not in the cards. Oh well, it is a lot cheaper than Harvard and a lot of those video lectures were prepared by top notch professors at places like Harvard and the University of Chicago. They claim having a professor from Harvard on video is far better than just any old professor in a classroom. Most of the instructors for technology courses are from industry and I am told that if you get one from the company you are interested in working for and do a good job, you are more likely to get a job offer in the future. Courses in other areas seem to be mostly those tapes and automation. They require a computer joystick for the educational software packages, so the school cannot be too bad, plus major Hollywood studios produce their multimedia software.

It does worry me that their tuition jumped by 20% in our area as soon as our local community college went out of business. I did not realize their tuition was geographically dependent. Their software costs are quite high since each course uses unique packages, including the ebooks generated by the professors. These materials seem be undergoing constant revision, but I suspect that is so the prior year's material cannot be sold in a secondary market among the students. Even though the average course size is one thousand students, eU does have these small discussion sections of 50 to 100 students run by the course graders. So, at least you can get help when you need it. Still, some courses use automated graders and I am not clear how that works, yet.

I was told the compositions in the first writing course and the programs in the first computer course are completely graded by the computer without the need for any human to look at them. An intelligent system not only designs the exam so that every exam is unique to every student in the course, but also uses your past performance profile to tailor the exam to your performance level. This allows even C students to get high point scores so they can feel good about themselves and show good results to their parents, who are probably financing their studies. Students are classified as Outstanding, Above Average, or Average, and then receive grades within those categories. Everyone has a chance to get a lot of A grades.

They sent me this funny form with their acceptance letter, where I must promise to not divulge any of my experiences in courses to any data collection process not approved by eU.com or they can deny me any future access to my records and rescind my degree. I don't understand the reason for that one at all. Oh well, I have no real choice, given my situation.

Summary

In the first scenario above, the student will receive the same quality of education whether he studies on campus or at home. He will participate with a group of his peers and will establish a network of relationships to utilize throughout his career. He can also get to know his instructors and his fellow students well.

In the second scenario, the student will participate in a distance-learning program set up like a mass production process. This is a clear second choice, apparently forced on the student by circumstances and costs. This option sacrifices the quality of education for the ultimate efficiencies and mass delivery of courses.

The real variable, which will decide between these future alternatives, is whether higher education institutions integrate all their face-to-face students in the same communication environment, prioritizing collaboration for all students and re-warding faculty who introduce new technology in this way. Regardless of what is written down, in most universities the rewards for faculty are inextricably linked with research and external funding and instruction needs only to be acceptable to obtain promotion and tenure. Innovation in education and exceptional teaching are not prioritized for young faculty at many institutions.

This is clearly a problem with administrations. While administrations focus their attention on competitive research and sponsored funds, the educational process is undergoing an unanticipated, unexamined, fundamental change. The next decade will bring some rude awakenings. Because of the time needed to change attitudes and bureaucratic processes, some of these awakenings may occur too late. Competition in instruction on an international and national basis will become the principle determinant of institutional success or failure in the next decade. We are entering a free marketplace era for the enterprise of education at the university level. The Web is the first communication system where consumer reaction to experiences with alternatives is cheap and easy to collect, organize and provide. One of the key premises underlying the concept of a free market is the free flow of relevant information and that is going to happen for individual courses, as well as degree programs.

The most important factors for future success will be the quality and talent of the instructors and their commitment to excellence in learning. Many institutions may well have to reassess the relative imbalance in faculty rewards between teaching and research. In addition, marketplace mechanisms will make the quality of teaching more visible to the public and prospective students. We can expect the threshold for acceptable teaching quality to rise. Regular students will opt for distance participation in some of their courses, not only because it is convenient, but also because they perceive no loss of quality. As long as both versions of the courses utilize the same technology and learning methodology this is going to be true.

Ultimately, the fundamental changes that could ensure the future success of university and college level institutions may well have to come from the accreditation agencies in realizing that we are evolving in a competitive marketplace and that it is their role to ensure that the consumer has access to the information needed to make fair market decisions. Ensuring that courses in accredited degree programs can always be transferred among accredited institutions and that accreditation might have to be assigned to individual faculty as well as individual degree programs will most likely be a part of that evolutionary process.

References

Cho, H. K., Turoff, M. and Hiltz, S. R. (2003, January). The impacts of Delphi communication structure on small and medium sized asynchronous groups. In *HICSS Proceedings*, IEEE Press.

Discenza, R., Howard, C. and Schenk, K. (eds.). (2002). *The Design and Management of Effective Distance Learning Programs*. Hershey, PA: Idea Group Publishing.

Gray, S. (2003, August). Moving — elearning vendors take aim in the changing environment. *Syllabus,* 16(1), 28-31.

Harasim, L., Hiltz, R., Teles, L. and Turoff, M. (1995). *Learning Networks: A Field Guide to Teaching and Learning Online.* MIT Press.

Hiltz, S. R. (1993). Correlates of learning in a virtual classroom. *International Journal Of Man-Machine Studies*, 39, 71-98.

Hiltz, S. R. (1994). *The Virtual Classroom: Learning Without Limits via Computer Networks, Human Computer Interaction Series.* Intellect Press.

Hiltz, S. R. and Turoff, M. (1993). *The Network Nation: Human Communication via Computer.* MIT Press (original edition 1978).

Hiltz, S. R. and Turoff, M. (2002, April). What makes learning networks effective. *Communications of the ACM*, 56-59.

Hiltz, S. R. and Wellman, B. (1997, September). Asynchronous learning networks as a virtual classroom. *Communications of the ACM*, 40(9), 44-49.

Howard, C. and Discenza, R. (1996). *A Typology for Distance Learning: Moving from a Batch to an On-line Educational Delivery System.* Paper to be presented at the Information Systems Educational Conference (ISECON) in St. Louis, Missouri, (October 1996).

Howard, C. and Discenza, R. (2001). The emergence of distance learning in higher education: A revised group decision support system typology with empirical results. In L. Lau (Ed.), *Distance Education: Emerging Trends and Issues.* Hershey, PA: Idea Group Publishing.

McIntyre, S. and Howard, C. (1994, November). Beyond lecture-test: Expanding focus of control in the classroom. *Journal of Education for Management Information Systems.*

Nelson, T. H. (1965). A file structure for the complex, the changing and the indeterminate. In *ACM 20th National Conference Proceedings* (pp. 84-99).

Turoff, M. (1995, April). A marketplace approach to the information highway. *Boardwatch Magazine.*

Turoff, M. (1996). Costs for the development of a virtual university. *Journal of Asynchronous Learning Networks*, 1(1).

Turoff, M. (1997, September). Virtuality. *Communications of ACM*, 40(9), 38-43.

Turoff, M. (1998, Spring). Alternative futures for distance learning: The force and the darkside. *Online Journal of Distance Learning Administration*, 1(1).

Turoff, M. (1999, January/March). Education, commerce, & communications: The era of Competition. *WebNet Journal: Internet Technologies, Applications & Issues*, 1(1), 22-31.

Turoff, M. and Hiltz, R. S. (1986). Remote learning: Technologies and opportunities. *Proceedings, World Conference on Continuing Engineering Education*.

Turoff, M. and Hiltz, R. S. (1995). Software design and the future of the virtual classroom. *Journal of Information Technology for Teacher Education*, 4(2), 197-215.

Turoff, M., Hiltz, R., Bieber, M., Rana, A. & Fjermestad, J. (1999). Collaborative Discourse Structures in Computer Mediated Group Communications. Reprinted in *Journal of Computer Mediated Communications on Persistent Conversation*, 4(4).

Endnotes

[1]   A great deal of recent evaluation studies are beginning to confirm our earlier findings based upon extensive and large scale studies at such places as SUNY, Drexel, Penn State and others. Some of these may be found in the Journal of ALN (http://www.aln.org) and on the ALN Evaluation Community web site (http://www.alnresearch.org).

Chapter II

Design Levels for Distance and Online Learning

Judith V. Boettcher
Designing for Learning, USA

Abstract

This chapter describes a multi-level design process for online and distance learning programs that builds on a philosophical base grounded in learning theory, instructional design, and the principles of the process of change. This chapter does the following: (1) describes a six-level design process promoting congruency and consistency at the institution, infrastructure, program, course, activity and assessment level; (2) describes a conceptual framework for designing online and distance learning programs; and (3) suggests a set of principles and questions derived from that framework. The principles are derived from the Vygotskian theory of cognition that focuses on four core elements of any teaching and learning experience — the learner, the faculty/teacher/mentor, the content /knowledge /skill to be acquired/or problem to be solved, and the environment or context within which the experience will occur. This chapter includes a set of principle-based questions for designing effective and efficient online and distance learning programs.

Introduction

What differentiates effective distance learning and online learning programs from those that are less effective, less efficient or less attractive to students? Do successful online and traditional programs share a common set of instructional design principles that might be more consistently applied?

This chapter describes a six-level design process that promotes congruency and consistency at the institution, infrastructure, program, course, learning, activity and assessment levels. This multi-level design process builds on a philosophical base grounded in learning theory and instructional design, as well as in the principles of change processes. The design process includes perspectives from a Life Style and Learning Style Design Framework (LS-TWO) that recognizes the influences of the life styles and learning styles of learners and faculty, and the challenges and power of the new technologies and their impact on communications and resources. It is hoped that the questions and principles derived from this framework will support instructional planners in the near term and also into the future. In summary, the goals and objectives of this chapter are to:

- Describe a six-level design process incorporating design at the institution, infrastructure, program, course, activity and assessment level.

- Describe a conceptual framework for designing online and distance learning programs.

- Suggest a set of principles and questions derived from that framework.

When a reader completes the chapter, they should have at their disposal a set of principles and questions for designing effective and efficient online and distance learning programs. These principles are derived from the Vygotskian theory of cognition that focuses on four core elements of any teaching and learning experience. Those four elements are: the learner, the faculty/teacher/mentor, the content /knowledge/skill to be acquired/or problem to be solved, and the environment or context within which the experience will occur.

Design Principle for Planning Distance and Online Learning

A fundamental principle for designing online and distance learning is that design happens not just at the course or program level by a faculty member. Achieving effective and efficient online and distance learning programs requires research, planning, and program design on many levels.

Institutions with a successful track record in this area often approach each degree, each certificate program, each outreach effort, as a distinct "product offering" for which a business plan must be researched and developed. For institutions wanting to get started in offering courses online or at a distance, this can be a good strategy, as it makes visible many of the unanticipated requirements and services for providing teaching and learning in virtual or distant spaces.

Designing online and distance learning programs on a broad scale is an institutional commitment, requiring design at a minimum of six levels corresponding to the *components* of teaching and learning experiences and to the *structure* of an institution's delivery systems. The next section briefly describes these six levels of design and provides lists of questions to help the design process at each level.

Six Levels of Design

Effective instructional design for online and distance learning usually requires instructional planning at these six levels:

- Institutional design — congruence with institutional mission
- Infrastructure design — management of and access to student services, faculty services, learning resource services
- Degree, curriculum, program, or certificate design
- Course design
- Unit and learning activity design
- Student assessment design

Effective articulation of the institutional and infrastructure design levels is a campus wide responsibility generally led by a group or committee composed of representatives of faculty, staff, students and administration. These two top-level designs are best served with the involvement of those representing the entire campus community and high-level leadership. Why is this multi-level design process recommended? Because it supports effective and efficient online and distance learning by incorporating an institution-wide vision for students and for the teaching and learning environment used by students and faculty. To work well, the institutional vision reflects a shared philosophy of teaching and learning.

A shared philosophy of teaching and learning cannot be assumed, nor can it be easily or quickly achieved. Fortunately, it is not necessary for design to be operationalized at the campus level to launch a single online or distance learning program. Many successful online and distance programs can be and are envisioned and designed at the college level and at the graduate professional level with only nominal support at the institutional level. However, any online or distance program requires close cooperation and visioning with the central infrastructure operations group.

The design approach for a college program is the same as the design at the institutional level; what changes is the unit of design and the analysis and processes for design. The primary goal is that of design congruence among goals, infrastructure, program, students, activities and evaluation.

Characteristics of the Design Process

A key characteristic of instructional design is its *iterative* nature. All systems are by their very nature interdependent with the other elements within the system. As systems change and as the environment within and about an institution changes, all elements of a system need to be systematically reviewed to be congruent and consistent with one another. This means that the timing of the design at each level must be planned for as well. Is design ever complete? Only temporarily.

Another key characteristic of design is *perspective*. Ideally, groups responsible for design of teaching and learning should share common philosophies and viewpoints. How we think about teaching and learning impacts the decisions we make regarding environment, expectations, resources and goals. This is why our institutional staff, including developers of content and systems should all be constantly learning, not just about technologies and systems, but also about teaching and learning. How we think impacts how we act and how we design.

Even decisions about the design of administrative systems cannot be done in isolation from the design of courses and degrees. For example, the process of applying to an institution, of being admitted to an institution, and queries about programs all become part of the total learning experience for students. All interactions between students and the institution are impacted by design and behind the design is the "code" which captures and, all too often, causes design to become rigid and to become its own law, however unintended (Lessig, 1999).

Ideally, the design at each of these six levels reflects philosophies of teaching and learning that are consistent with the institutional mission and consistent with the expectations of the students and society being served. Figure 1 summarizes these six levels of design, and identifies the group or individuals usually responsible for the design at that level. The length of time for a design cycle at each level is also suggested.

Six Levels of Design for Learning

The Lifestyle and Learning Style Design Framework (LS-TWO) that is the described later in this chapter suggests a philosophical basis for design based in constructivism and the learning theories of Vygotsky (Vygotsky, 1962, 1978). The principles suggested by this framework — or by any other chosen framework — inform all six levels of design.

The people involved in designing for teaching and learning at an institutional and infrastructure level include a wide range of institutional faculty and staff. Viewing design as a multi-level process keeps the process practical and realistic. Design decisions at the infrastructure and institutional level require consensus followed by time for design and implementation. Design decisions at the program and course level require a faculty team consensus. Some design decisions at the course, unit and assessment levels can be made more quickly, but must be made within the context and within the framework of the infrastructure design and resources design.

Level One: Institutional Design

The design work to be done at an institutional level is similar to strategic planning and positioning of an institution. A good clue to an institution's current strategic

Figure 1. Six Levels of Design

Six Levels of Design	Design Responsibility	Sponsor/Leader	Design and Review Cycle
Institution	Entire campus leadership and community	Provost, CIO and Vice-presidents	3-5 Years
Infrastructure	Campus and Technology Staff	Provost, CIO and Vice-presidents	2-3 Years
Degree, Program	College/Deans/Faculty	Dean and Chairs	1-3 Years
Course	Faculty	Dept. Chair	1-2 Years
Unit/Learning Activity	Faculty	Faculty and/or Faculty team	1-2 Years
Student Assessment	Faculty	Faculty and/or Faculty team	1-2 Years

positioning is completing an association test. For example, what type of student comes to mind when MIT is mentioned? Focused, talented engineering and technical students. What type of learning experience comes to mind when Oberlin College is mentioned? Liberal arts colloquia with active, involved students. When the Open University of the UK is mentioned? Working professionals doing online and distance learning supplemented by tutors. Or Penn State University? Large, land grant institution with a wide variety of applied sciences and disciplines — and yes, a great football team! An institutional design includes completing this association test, both for what the associations are now and what an institution wants them to be in the future.

Institutional planning generally begins with an institution's current vision and mission statements and then proceeds through a data collection and input process that addresses a set of questions such as the following:

Institutional questions:

- What is our institution's primary mission?
- What students are we planning to serve?
- To what societal needs and goals is our institution attempting to respond?
- What life goals are most of our students working to achieve?
- What programs and services comprise our primary mission? For whom?
- What type of learning experiences are our students searching for? And what types of learning experiences do we want to provide?
- What are we known for at the current time? And what do we want to be known for in the next five years? The next 10 years?
- What are the strengths of our mature faculty? Of our young faculty?

If an institution is in the planning cycle for programs in online and distance learning, the institutional design process includes questions such as the following.

Institutional questions focusing on online and distance courses:

- What is the institution's catalyst for offering online and distance courses? Or for modifying current programs to include online components?
- Is our primary or secondary student population requesting these kinds of courses and programs? Or this type of online interaction?
- Are online and distance programs part of a longer term strategic positioning of our institution? Is this where we want to be with one-two degree programs? Or many?

- What changes in our infrastructure are needed to match our desired services, programs and students?
- Does our institution have any special core competencies, resources, or missions that are unique regionally or nationally that might form the basis for unique and specialized online and distance programming?

Some primarily traditional institutions have successfully expanded beyond their current "branding" with online and distance learning. Penn State, for example, launched their World Campus Project in 1997, building on their expertise with their extensive distance learning outreach programs. Another well-established distance-education provider, the University of Maryland University College (UMUC) is now a leader in online learning, as well. As of mid-2002, UMUC was offering 15 bachelor's and 17 master's degrees fully online, as well as almost 50 online certificate programs. The University of Florida has been successful with professional degree offerings, combining synchronous and asynchronous course experiences with mentors at various physical locations (Riffee, 2003). For institutions with experience in offering distance programs, going "online" has meant adapting current processes and programs to new media, and to new applications for interactions.

Other institutions that are launching themselves as only online or distance institutions include Capella University, and Jones International University. This is a more difficult task. Many institutions launched as virtual for-profit institutions were launched with great fanfare and money and then quietly closed (Arnone, 2002; Blumenstyk, 2001; Boettcher, 2002).

Meanwhile, the Open University of the United Kingdom, a large, single-mode institution launched in the early 1960s, is rapidly expanding the number of courses with online components for delivery in the UK and elsewhere. And the number of students accessing information via the Internet is growing rapidly.

A guiding principle emerging from these institutional efforts is a solid business principle of staying close to your core competencies, which also means designing to your strengths. Let's take a closer look at how the Open University is strategically well integrated — balancing its mission, vision, planning, development and services.

Open University System Model — Institutional Design Example

The Open University in the UK is well known for its very successful distance learning programming which now reaches over 200,000 students, representing 35% of all part-time higher education students in the UK (http://www.open.ac.uk/). According to John S. Daniel (1999), the Open University owes much of its success to a focus on the following elements of their teaching and learning program:

- Well-designed multimedia teaching materials
- Personal academic support for each student
- Efficient logistics
- Faculty who also conduct research

These four characteristics reflect the priorities and goals of the Open University. These goals influence the design of their institutional structure, their faculty and staff, and their institutional priorities. The Open University invests heavily — by virtue of faculty, developers and time in the design and development of effective multimedia teaching materials. Their system of Associate Lecturers (tutors) includes a program of substantial and extensive preparation for the tutors. Evaluation of courses, materials and tutors are part of every course and program. To ensure students are well supported, help is available 12 hours a day, every day. And to ensure the continued development of effective materials, faculty conduct research into learning that "continues to enhance the excellence of teaching materials" (OU Strategic Plan, 2002-2012).

The Open University has apparently been successful with their focus on developing excellent materials, logistics and support. "External assessment of Open University teaching standards has awarded 'excellent' ratings to 18 of the 25 subjects assessed at the end of 2001."

The principle that this example demonstrates is that all components of an institution should reflect the vision and the mission of that institution. A good overall guiding question for creating strategic plans for institutions and for evaluating initiatives is, "Does this service, program, or initiative support the overall mission and overall vision of our institution?"

This question is helpful for transitioning our thinking to the design level for the infrastructure. People often think that buildings, classrooms, Web applications, communication services and servers are neutral as far as having an effect on teaching and learning. Nothing could be more misleading.

Level Two: Infrastructure Design

Design of the infrastructure includes design of all the elements of the environment that impact the teaching and learning experiences of faculty and students and the staff supporting these experiences.

It includes design of the following:

- Student services, faculty services, and learning resources
- Design of administrative services, including admission processes, financial processes and institutional community life events
- Design of physical spaces for *program launching* events, hands-on, lab or network *gathering* events, and also *celebratory* graduation events

Physical and Digital Plants

Infrastructure design for online and distance teaching and learning programs focuses on the design of the Internet and Web infrastructure. A useful way to think about the online infrastructure is to think in terms of analogs to the traditional elements of physical plants on campuses. For example, infrastructures for online learning have offices, classrooms, libraries, and gathering spaces for the delivery and management of learning and teaching. However, these offices and classrooms are accessed through Web services, rather than through physical buildings.

The good news about online infrastructures is that they support an unparalleled new responsiveness, feedback, and access for learning activities. For example, e-mail is definitely speedier and more satisfying than mail or phone tag. The bad news is that the "online campus" — just like the traditional campus — depends on a complex, costly, ever expanding "online campus" resource. The need for new administrative structures and processes now involves more efficient networks, servers, applications, and services. Expectations of speedy and accurate 24 hours a day/seven days a week (24/7) response is growing almost as fast as the technologies are growing.

Certainly expectations are growing faster than people can change and adapt the services. Expectations of faculty response have reached the point where some institutions are instituting policies that a faculty member will respond to e-mail within 24 or 48 hours during the normal workweek. This type of requirement can impose significant changes in faculty work habits and research focuses.

After almost 10 years of building online campuses, we now know that a "digital plant" infrastructure is needed to support the new flexible online and distance environments. We know that this new digital plant needs to be designed, built, planned, maintained, and staffed. The infrastructure to support the new programs cannot be done with what some have called "budget dust" (McCredie, 2000). It is not nearly as easy or inexpensive as we all first thought. Some gurus suggest that a full implementation of a plan for technology support on campus "costs about the same as support of a library — approximately 5% of the education and general budget" (Brown, 2000).

Components of a Digital Infrastructure

What exactly is a digital plant infrastructure? One way of describing this infrastructure is to think of it in four major categories of software, hardware and people. These categories are:

- Personal communication tools and applications,

- A network that provides access to Web applications and resources and access to remote, national and global networks,

- Dedicated servers and software applications that manage campus services. (Remember that servers are just specialized computers dedicated to "serving" or providing focused services.) These servers support Web services, such as in-going and outgoing mail, web sites, Web applications, campus directories, program and course management systems, administrative services such as financial, student services, and human resources, and the new e-commerce servers, and

- Software applications and services from external providers, such as research and library services, some of which are licensed to the institutional community, Internet services, and out-sourced services, such as network services, etc.

None of these systems work without people to manage them, so a key component of the digital infrastructure is the group of individuals who make the systems work. Figure 2 shows the people at the center of this infrastructure (Boettcher and Kumar, 2000).

Higher education institutions have been designing and building digital infrastructures for teaching, learning and research since the early 1990s. (The building of networks for communications goes back to the early 1980s.) The period of design and development of these teaching and learning digital infrastructures is now well into its second and even third generation. Internet2, a consortium of 202 universities working in partnership with industry and government, is leading work on the next generation of advanced networks and applications (see http://www.internet2.edu/).

Some of the questions that might be used to guide the development of the digital infrastructure are listed below. The first set of questions addresses infrastructure design questions for long-term planning. These design questions are part of the strategic planning processes and are addressed in collaboration with institutional planning groups. The next set of questions maps to the four categories of the digital plant.

Figure 2.   Teaching and Learning Infrastructure — "Digital Plant"

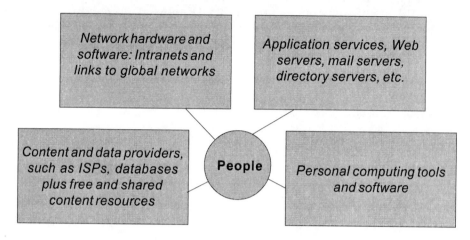

Institutional Infrastructure Level Questions for Long-Term Planning

- What are our current institutional infrastructure strengths? Weaknesses?
- What changes in our infrastructure are needed to provide existing needs and to plan for future services, programs and students?
- Is the infrastructure scalable? Can it accommodate our expected growth targets?
- Is the infrastructure reliable? Are backups and alternative resources in place to support mission-critical applications and services?
- Is the infrastructure secure enough for our needs and services?

Infrastructure Questions Mapping to Four Categories of Digital Plant

Personal communication tools and applications:

- What are the tools that we expect each of our students to own or to have access to? Do we expect students all to be proficient at word processing? At e-mail? At Web applications? At researching on the Web? At using collaborative tools? At using one or more course management systems?
- Will all students have their own computer? Their own laptop?
- How many hours a week do we expect students to be working "online"?

- How often and when do we expect students to require technical support?
- How often and when do we expect faculty and staff to require technical support? What type of support will they need?
- How will faculty and staff become proficient at the use of the systems?

Networks that provide access to Web applications and resources and to remote, national and global networks:

- What physical networks are needed to support Web applications, such as e-mail servers, directory servers and Web application services?
- Do we need or want wireless services in any of the physical spaces?
- What speed will be needed for which services?
- How often will the higher bandwidths be needed? Will higher bandwidth services be needed for video conferencing for programs? For meetings? For downloading large files? For streaming video?

Dedicated servers and software applications that manage campus services:

- What types of interactive Web services will be provided? What hardware and software will be required?
- What type of course or learning management system will we use?
- What type of administrative systems will we use?
- Which group/organization will design/develop the Web presence for the institution?
- Will a portal approach be used?
- What do we need to do to assure student, faculty and staff accessibility, and self-service from anywhere at anytime?
- How will these applications and servers be secured?
- What type of authentication mechanisms will be used?

Software applications and services from external providers, such as research and library services that are licensed to the institutional community, Internet services, and out-sourced services, such as network services:

- What licensed services are required and desired?
- How will research and access to resources be provided?

- What services are to be available through consortia, etc.?

- What budget is required to support these services currently and into the future?

Technology decisions for students have always been part of the planning design process for distance learning and, obviously, they are even more important today. A comforting way of thinking about the technology for the infrastructure design level is in terms of the generations of technologies used in distance learning (Sherron and Boettcher, 1997). Distance learning was made possible with the widespread availability of technologies, such as the mail, radio, telephone, television, and audio and videocassettes. In the 21st century we simply have more technology and more choices. But, as noted before, expectations of service seems to be growing faster than the technology implementations.

Before moving on to the design questions at the program and course levels, a brief note about change might be useful.

Role of Change in the Institutional and Infrastructure Design Levels

The design processes at the institutional and infrastructure design levels inevitably create expectations about how to manage institutional change that may result from the design planning process. Implementing any institutional or infrastructure design plan often means a realignment of priorities and changes in processes. A body of change literature is available to help with the process of design and the communication and community-wide activities to support change. Two articles are particularly to be recommended for dealing with strategic planning, goal setting and change management (Kaufman and Lick, 2000; Lick and Kaufman, 2000).

Now let's look at the design of programs and courses. Design issues at these levels are principally the responsibility of the institutional academic leadership, assuming that the infrastructure design is in place.

Level Three: Program Design

Design at the program level is basically curriculum design and focuses on which degrees, programs, and certificates are to be offered to students in the online or distance environment. Online and distance learners generally are interested in achieving or completing an instructional goal that can assist in their current or future career path. The expectations of the use of online programs for lifelong learning for leisure and personal interest will probably emerge over the next decade or so, but the majority of online learners want a program of learning that supports their career goals. At Florida State University in the mid-90s, one of the foremost recommendations of an institutional distance learning committee was to launch new online degree programs. That committee recommended investment in degree programs, such as master's degrees in Information Studies, Social Work, Communications, and Criminal Justice.

At the program level of design, instructional planners answer questions about the type of program to be offered, to whom, and over what period of time and at what cost. While teaching seminars on designing and developing online and distance programs, it was satisfying to see how relieved the seminar participants generally were to learn the importance of focus, pilots, and phased implementations. Some participants have come to such seminars with instructions to "find out how to put all the college programs online!" Participants left the seminar generally pleased with two recommendations: (1) Focus on those programs where they had core competencies and strengths; and (2) Plan a phased approach, gaining experience in delivering programs in one or two areas before launching into many others.

As mentioned before, the design of every program should include a business and marketing needs analysis. Before launching a distance learning degree program, administrators like to be assured that there is a market for that particular program. The fully online institutions, such as Capella University and Jones International University, initially focused on programs offering master's degrees for working professionals in areas such as business, business communications, education and information systems, and then branched out to undergraduate completion programs in these same areas. Both of these institutions are fully accredited by the Higher Learning Commission, a member of the North Central Association, an accrediting body for institutions of higher education in the United States.

Program level planning questions:

The first set of decisions for program planning can be guided by the following sets of questions.

Curriculum questions:

- What is the degree or certificate program to be offered online? Will it be a full master's degree (10 to 16 courses), an undergraduate minor (four to six courses) or a certificate program (two to four courses)?
- Which college or department will be offering the program?

- Is the program currently being offered in a traditional distance or campus format, or is the program a new degree or certificate development effort?

- What types of courses are envisioned? Will these courses be a fully developed "course in a box" with a minimal amount of interaction or a highly interactive and collaborative course requiring or using many online resources and applications? Will the courses be Web courses, Web-centric courses, or Web-enhanced courses? (Boettcher, 1999)

Design and development questions:

- Who are the faculty who will design, develop and deliver the courses in the program?

- Who will lead the effort to develop the degree or certificate program for online or distance delivery? Will there be a project manager? (This is highly recommended.)

- What course management system or similar Web tool will be used for the course management? What tools will be available and supported for the interaction and collaboration activities?

- What types of teaching and learning strategies will we use?

- What is the schedule for design and development? What is the schedule for the delivery of courses?

- Which organization will be marketing the program? What is the schedule for the design and development of marketing materials?

- What resources and tools are envisioned for the library and research materials?

- What is the relationship between the academic developers and the technical support personnel?

Faculty questions:

- What training will be available to faculty as they transition to online teaching and learning programs?

- What tools and resources will be available to faculty?

- What tools and resources will faculty be expected to use?

- What support will be available to faculty, especially for the delivery of the first one or two courses?

- Will faculty have any released time or budget for teaching and learning resources in this new environment?

Student questions:

- Who are the students who will enroll in this course of study? How will we find them and market the program to them?

- What will our students bring to the program experience?

- What tools and resources will the student in this program or certificate program require or be likely to use?

- Where will the students be doing their learning? What level of network access is required or recommended?

This list of questions is intended as a catalyst for a college or department. As the process gets underway, many more questions will emerge that will be specific to the campus/institutional environment and culture.

ACTIONS Planning Model (Bates)

In this area of rapid technology development, an essential early question at the program level of design is just what set of technologies faculty and students will be using for their teaching and learning experiences? A planning framework for student technologies was proposed by Tony Bates, one of the founding members of the Open University staff in the United Kingdom in early 1990s (Bates, 1995). The ACTIONS acronym is an easy way to remember the key elements of the decisions for technologies. It is not only useful, but also highly recommended that this framework also be applied to the faculty who will be teaching and learning with the students in these programs.

The ACTIONS model provides a practical decision-making framework for identifying and selecting the set of technologies to which students can be assumed to have access during the course of a program. These questions can effectively be used at the beginning of the design of a program, and also as a checkpoint during the development process. The seven questions follow, with updated annotations.

(A) Access: How accessible is the technology for the students?

The question of technology access was particularly important in the mid-1990s, when technology access by students was relatively modest. Now, students are very likely to have sufficient communications access and technologies and may be even more skilled at their use than faculty. Thus, the question has become which set of technologies is important and what is the expected number of hours of use and the speed of access to be recommended. The latest data from the Campus Computing Study of 2002 suggests that more than 75% of all students own their own desktop or notebook computer (Green, 2002). For online and distant programs, this figure is probably higher, as learners often have access at work, or shared access at home. If all students have their own computers and access to the Web, faculty/mentors and support staff can assume easy convenient access and often unlimited access to networks and services. This type of access greatly impacts the design of requirements, communication activities, and course activities and experiences.

(C) Costs: What is the cost structure of the technology?

Any assumed technology needs to be affordable for the three partners in a teaching and learning experience — the faculty, students, and institution. This is where the question of support and infrastructure needs intersect in design. Support for specific applications and tools needs to be considered in some detail.

(T) Teaching and learning: What kinds of teaching and learning experiences are required for the program goals?

The types of goals and objectives and the types of learning activities should be appropriate for the speed and convenience of access, as well as for the life style and learning styles of students and faculty.

(I) Interactivity and user-friendliness: What kind of interactivity is enabled? How easy is it to use?

Interaction between faculty and students, among students and between students and learning resources is fundamental to the dialogue nature of effective and enjoyable learning. The set of technologies that are selected and supported should be easy, friendly, and intuitive.

(O) Organizational issues: What are the organizational requirements and barriers to success?

Delivering degrees, programs, and courses to students online and in distance environments requires a supportive, always-on, reliable, and secure digital infrastructure. Systems and processes should be in place to recruit, admit, support, evaluate, advise students, and provide access to library resources. These support structures are critical to student learning and for faculty services.

(N) Novelty: How new is this technology?

Technology that is too new is often difficult to deploy and support effectively. So new technologies are best introduced in pilot modes within established contexts.

It is tempting to be ultra-conservative in the use of media, but older technologies and media bring their own set of serious support challenges.

The importance of the decision as to which set of media and technologies to use has been greatly oversimplified by the widespread availability of the Web. Design decisions still should weigh assumptions as to network speed and the ability to use Web tools, as well as audio, video and multimedia technologies. However, it is also true, that by the time a course is deployed, technology tools and environments can change as well. So the question is not what is the level of knowledge and access available today, but what is the level of knowledge and access likely to be when the course is delivered to a specific target population?

(S) Speed: How quickly can the courses be developed and delivered?

Quality courses can take a significant amount of time to design and develop. Development of a complete online or distance course can easily take 18 to 30 months by a team of faculty developers, even with the better tools that are now available. On the other hand, adding Web components to existing distance courses can be designed and developed more quickly. How quickly can a course be developed and delivered? It depends on the design decisions that are made (Boettcher, 2000).

Another piece of good news for developing courses is that there is now a rich body of content available in digital format. The tools for identifying, selecting, organizing, and structuring content are also rapidly evolving. The time to prepare for course delivery has the potential of improving if existing digital content can be redirected toward instructional experiences. Some of the coming design challenges are to improve the cost/benefit ratio and the "fitting"-ness of the instructional program to the whole person.

Faculty Technologies and Support Needs

Faculty also require support in adapting to the new online and distance environments. When planning for faculty support needs and training, institutions may find the following survey of interest. An informal Web survey in the Fall of 1999 asked faculty development personnel at 20 universities to identify the top three issues that their faculty were facing in the development of their digital teaching and learning infrastructure. The six most frequently cited issues were: (1) the need for institutional support for the use of the new teaching and learning tools and services; (2) support for planning the new teaching and learning experiences online; (3) faculty training; (4) funds; (5) governance; and (6) a concern for quality.

The survey also asked which services faculty most frequently requested. These results were both surprising and insightful. Faculty were asking for help in what we now call "course life-cycling" for Web content. They were also asking for help in: (1) course creation planning; (2) developing of courses; and (3) managing the delivery of Web courses. Also mentioned were help in archiving and data mining of courses that had already been delivered.

Faculty also expressed a desire for more support for learning general productivity software applications and for learning course management and development tools, such as mailing lists, spreadsheets, and integrating video components into their Web courses. Finally, faculty frequently mentioned support for students. Just as faculty require support for shifting their teaching and learning habits, many students, especially if they are more than 30, grew up after the digital convergence. Such students have ingrained habits from the traditional campus teaching and learning models.

Course Design — Level Four

Now let's look at the level of design at the course level. This is the level that is assumed to be the total responsibility of the faculty member. Is that assumption true for the design of online and distance courses?

In online and distance courses, the independent or stand-alone course is the exception, rather than the rule. Most online and distance courses are part of a curriculum, program, certificate or degree program. This means that course level design occurs within the context of the larger degree or certificate program and probably in collaboration with other faculty and the academic department.

The downside of this contextual dependency is that design is likely to be more complex, as it is a team effort, rather than an individual effort. Design is also more likely to be "unbundled" from the development phase of the planning cycle. This is good, but presents a different process from the bundled approach. The benefits of this dependency means that the course design process is collaborative and that faculty can benefit from the design work at the program or degree level. The set of technologies and assumptions made at the program and infrastructure level will likely be similar to those at the course level. The same is true for the Web applications, and the interactive and collaborative tools that a course will use, including the database or sources of resources to be used. All of these types of course design questions are typically completed in the context of the infrastructure discussed earlier.

What design decisions should be made by the course designer in the online and distance environment? The course designer is primarily responsible for questions on the course content, objectives, student goals, learning experiences, and assessment for a particular course. Many of these questions parallel questions at the program level design.

Course questions:

- Where does this course fit within the context of the degree or certificate program to be offered online? Is it an early course focused on discipline concepts, peer discussions, and standard problems or a later course focusing on applications and complex scenarios?

- What is the core set of knowledge, skills, and attitudes/values expected to be acquired by the students in this course?

- Is this course currently being offered in face-to-face or distance mode now, or will this be a new development effort?

- What types of learning activities are envisioned? Will students be learning concept-dense content or discussing and solving problems collaboratively? Or both? Are the learning strategies specifically recommended and suited to these goals for this content?

- What tools will students need for completing the learning experiences designed for the course?

- What is the set of content resources *required* for this course? *Recommended* for this course? What content resources will students use to *customize* the learning experience for their needs and state of knowledge and desire?

- Will students be a cohesive cohort group? Or will the students in every class generally be new to one another?

- Will students be familiar with all the technologies, tools, and data resources? Are there tools or applications specifically recommended for the content of this course?

Design and development questions:

- What faculty member is the primary designer/deliverer of this course? Or, is it a team of faculty members?

- What faculty member will lead the effort to design, develop and then deliver this course over time? Will it be one faculty member or a team?

- What course management system or similar Web tool will be used for the content management? For the interaction and collaboration activities?

- What is the schedule for design and development and delivery of this course?

- What resources and tools are envisioned for the library and research materials?

Faculty questions:

- What training will be available to faculty as they make a transition to online teaching and learning programs?

- What tools and resources will be available to faculty?

- What tools and resources will faculty be expected to use?

- What support will be available to faculty, especially for the delivery of the first one or two courses?

- Will faculty have any released time or budget for teaching and learning resources in this new environment?

- What is the state of the faculty's personal teaching and learning technology? What is the set of technologies and tools to be used by the faculty for this course?

- Will faculty have access to printers, databases, and cameras? Will faculty need these tools?

- What type of access to the network is recommended and available? Will dial-up be sufficient, or will DSL or cable access be recommended or required?

Student questions:

- Who are the students taking this course? What are their hopes and expectations? What future courses will depend on the knowledge, skills and attitudes acquired in this course?

- What knowledge and expertise do the students bring to the course? What is their zone of proximal development (Vygotsky, 1978)?

- What life experiences will the students bring to the program experience?

- What tools and resources will the students in this program or certificate program require or be likely to use?

- What types of teaching and learning strategies best suit the students in this course? What are the life style and learning styles of the students?

- When and where will the students be likely to gather for their collaborative work? When and where will they do their more self-study activities?

- Where will the students be doing their learning? What level of network access is required or recommended?

While the number of design questions can seem overwhelming initially, much of this design work at the course level is very familiar to faculty. Faculty are often unconsciously competent at instructional design. They have often been practicing these skills for years, making preliminary assumptions about content, students' goals, students' knowledge, and environments during an informal course planning process and then refining it after meeting with a group of students. The key difference in applying these skills to online and distance learning is these courses are designed and developed in advance of meeting with students and independent of any particular group of students. Thus, the more intense design process. Good course design makes for effective learning by students, effective teaching and mentoring by faculty, more satisfied students and faculty, and overall positive faculty/student interactions and experiences.

Instructional design of complex learning for complex individuals is neither simple nor straightforward. However, this set of instructional design questions can help to create more effective and efficient learning experiences and environments.

The next two design levels are within the course parameters and generally are the responsibility of the course faculty/mentor.

Level Five: Unit/Learning Activity

Design questions and strategies specific to the unit/learning activity level and the assessment level are based on a Life Style and Learning Style Framework (LS-TWO) and the four knowledge bases within the framework. This framework focuses on integrating analyses of student life-style and learning styles into instructional planning. The framework is grounded in four knowledge bases: (1) traditional and current instructional design principles; (2) insights and principles from current cognitive research; (3) the dynamic nature of content resources; and (4) teaching and learning environments enabled by ubiquitous and mobile computing.

The Life Style and Learning Style Design Framework (LS-TWO) integrates traditional questions on program planning with individualized life and learning styles. The design principles and questions derived from this framework include:

(1) external planning and service characteristics from the institutional, infrastructure, and curriculum design levels and (2) planning for the internal cognitive processes and knowledge construction by students.

Examples of cognitive learning style design questions include: "How do students process information?"; "What do students construct in their minds when challenged with new concepts and rich content structures?"; "What knowledge do students bring to the learning experience?" and "What types of collaboration activities support learning?"

Another unique characteristic of this framework is an emphasis on the life-style (LS) of the learner. The life-style of the learner includes all the elements in a learner's current life situation. We often forget to consider the finer details of the life environments of faculty and learners when designing learning experiences.

Where will the learner be working? Will they have a personal space where they can control sound, temperature, disturbances, and network access? Will they have to "ask" their family if they can use the phone line for 15 minutes or an hour? The Life Style focus of the LS-TWO Design Framework encourages analysis of the where, when, with whom, and with what resources the learner going to be doing their learning work. This consists of constructing his/her own new knowledge and applying knowledge and solving problems.

The array of new mobile, wireless technologies enables learners to study anywhere at any time. Initially, the ability to study anywhere seemed to hold the promise of solving many problems associated with access to learning. We have not addressed the question of just when and just where this "anytime" is likely to occur, however. Many of us do our teaching and learning "anywhile." While we are driving, while we are talking on the phone, while we are cooking, and while we are talking to family — perhaps even while we should be sleeping or spending time with the family. Time is a scarce commodity and is getting scarcer.

One of the foundations of the framework is traditional instructional design integrated with Vygotsky's learning theories. Let's look at the primary principles of Vygotsky's theory and consider how instructional planning can benefit from an integration of Vygotsky's ideas with traditional design theory. We will then see what principles and questions we can derive from this integration of ideas.

Vygotsky's View of Teaching and Learning

There are at least three major concepts within Vygotsky's theory that provide principles for designing instruction. These concepts are:

- Viewing the teaching and learning process as having four major components — learner, faculty mentor, problem or content to be learned and the context for the teaching and learning experience

- The concept of the zone of proximal development
- The process path of concept development

Four Major Components of Teaching and Learning

Let's start with the first major concept. Any structured learning experience has four components — the learner (Le), the faculty mentor (M), the knowledge/skill (K) or attitude to be learned, and the environment (E) in which the learning is to take place. This view of the learning process provides an easily remembered four-point checklist (LeMKE) for designing instructional experiences.

The Learner (Le) Variable

The learner as a key variable in instructional design is well acknowledged. Instructional design asks questions about the learner, such as: what does the learner know now and what does the learner bring to the experience (prior knowledge, both accurate and inaccurate)? And what is it that the learner hopes to learn or achieve by participating in the teaching and learning activity?

Traditional instructional design encourages a thoughtful analysis of learners and an estimation of the student's knowledge, but constraints often dictate that the learner is viewed as a member of a population with many similar characteristics. Knowing what we know about learners, i.e., that all learners are likely to "know differently," suggests that we design to a variety of backgrounds and expectations. In other words, designing for a variety of knowledge constructs and accommodating a variety of content resources and activities.

Principles and questions for the learner:

What are some learner principles relating to life-style and learning style? What are some of the questions that we want to ask about our learners' state of knowledge?

Learner Principle #1. Adult learners bring personalized/customized knowledge to the learning experience. They bring their particular and personal zone of proximal development.

Questions:

- How do we determine a learner's zone of proximal development? What is the learner ready to learn? What is already known? What is partially known?

- What knowledge base about this content or closely related content does the student currently have?

- What kinds of problems can students solve now? What kinds of problems do we want students to be able to solve at the conclusion of the experience?

- What instructional experiences will facilitate the constructive learning work in a person's brain?

- What experiences can lead adults to contribute their perspective and opinions?

Learner Principle #2. Adult learners are very constrained in terms of time. Yet, learning takes time. Everything else being equal, more time spent on learning tasks generally equates to more learning.

Questions:

- What is it that learners want to know and need to know?

- How do we customize/personalize the content?

- What is the core conceptual knowledge to be learned? What are the best set of tools and applications for learning the core conceptual knowledge?

- What type of application knowledge is required?

Learner Principle #3. A learner's life-style impacts that learner's time on task and focus.

Questions:

- Where will the student be doing his/her learning? What time of day?

- What tools will be readily at hand?

- What resources for learning will be available?

- When and where will the learner meet/interact with other learners?

- What is the right mix of social interaction/learning for this learner?

Learner Principle #4. Adult learners bring personal and professional goals to the learning experience.

Questions:

- What instructional experiences can facilitate the reaching of the goals?

- What is the best or preferred mix of learning experiences for a learner?

- What types of activities and what types of problems do learners want to be able to analyze and to solve? What do learners want to be competent at?

The Faculty Mentor (M) Variable

What about the faculty mentor variable in the teaching and learning experience? The movement emphasizing learner-centered teaching suggests, I think, that we need a new word for the human who structures, designs, leads, and supports learning experiences. We seem to vacillate between terms, such as faculty, teacher, guide, mentor, tutor, and coach — trying to accommodate the new learning philosophies that it is the learner who is doing the learning. For this document, the term faculty mentor has been used. (Combining designer and mentor did not work, as it resulted in something like dementor.)

What is the role of the faculty mentor? What do they do? In the traditional bundled model, faculty mentors design instruction at the level of the course. Faculty mentors also develop and deliver the course. In traditional campus models and in some online programs, it is the faculty mentor who defines the parameters of a course, determines "what is to be learned," and determines the structure and content of the course. Faculty mentors write, select and assemble materials and learning resources. They design, select, and present experiences for the learner. The faculty mentor also manages the delivery of the course, including the daily interactions with the students and assesses student learning.

In online and distance learning models, these roles are often unbundled into the distinct roles of designers, developers, and tutors. Many distance learning and online programs have two different faculty mentors: the faculty who is the primary designer/developer of the content and the faculty member who serves as the learning assistant or tutor who interacts online with students and who may serve as the face-to-face contact for learning (Riffee, 2003).

Principles/questions for the faculty mentor:

Here are the key core principles for the faculty mentor and associated questions.

Faculty Mentor Principle #1. Faculty mentors have the responsibility of designing and structuring the course experience for the learners.

Questions:

- Is the faculty mentor experienced at the process of designing and structuring the course experience for the particular student group and environment?

- What type of support might the faculty mentor require? Does the faculty mentor have the tools, the knowledge, and the resources to design and structure the course experience?

- How much of the course is prepackaged? How much is to be developed by the faculty mentor?

Faculty Mentor Principle #2. Faculty mentors generally rely on a fairly small set of familiar teaching and learning strategies.

Questions:

- What are the teaching and learning strategies that are a good fit for the content and for the knowledge, skills, and attitudes to be learned by the students?

- What are the types of learning experiences that best fit the learners and best fit the knowledge and experience of the faculty mentor?

- Are there other teaching and learning strategies that the faculty mentor may want to develop for teaching this type of content?

Faculty Mentor Principle #3. Faculty mentors have a limited amount of time for the design, development and delivery of a course. Can a quality course be achieved within the constraints of time and support? Are there additional support personnel?

Questions:

- Where will the faculty mentor be meeting with the learners? How often? How long will the online or face-to-face meetings be?

- How much interaction is expected of the faculty mentor? How much preparation? What are the tools for interaction and communication?

- How will the faculty mentor interact with the institution providing credit?
- What resources and tools will be available for the management of the course?
- What resources might be designed, developed, and used within a consortium?

The principle of limited time reminds us that just as time on task is an important determinant in learning, time spent in interacting with students or monitoring a student's learning experience can be an important determinant in the quality of the overall experience. Yet, time is a cost. This is only one of the challenges of design. We have become accustomed to talking about a parent's quality time with children. Perhaps we should also become more aware of a faculty's quality time with students and the tools needed to support that quality interaction. A well-developed infrastructure that is easy for faculty to use can be a factor in achieving quality time on task for both learners and faculty mentors.

The Knowledge (K), Skill, Attitude to be Learned or Problem to be Solved Variable

In all learning theories, the task of learners is to acquire the knowledge, skill, and attitudes that are needed or desired. Vygotsky's theory leans towards the use of what we now call problem-based learning (PBL). I think it is noteworthy that many of the curricular innovations of the 1990s are based on problem-based learning. The studio physics program designed by Jack Wilson at Renselaaer in 1990 is one of the better-known course redesign efforts based on PBL (Kolb, 2000). For instructional design purposes, the content of a course can be any or all of the following: a body of knowledge and concepts to acquire, a set of skills to be acquired and practiced, and attitudes and values towards the knowledge and the skills.

The question of just what it is that we hope learners will take away from an instructional experience is a question that may be changing soon. We may want to divide the traditional instructional design question into two parts. Rather than asking what content is to be learned, we may want to ask what knowledge needs to be in the learner's heads and what knowledge does the user need to be able to find and to use external knowledge bases effectively?

This new design question also impacts the structure and types of content resources. We now structure courses to deliver all the knowledge that a learner might need, "just in case." But, we are moving towards a "just in time" model. We have seen this shift in manufacturing industries. The shift is coming for our information-based professions, as well. Knowledge keeps growing, expanding

— like the universe — and we can't know everything. But, "everything" can be quickly available. Thus, what is it that a child born in 2000 needs to learn to be an effective 21st century adult?

Principles/questions for the knowledge/content variable:

What does the life style and learning style framework suggest we ask about the content to be learned and about the problems to be solved for learners?

Knowledge Principle #1. All learners do not need to learn all content. All learners do need to learn the core or base concepts and to develop useful knowledge.

Questions:

- What content resources will be available for the faculty and for the students?
- Will the core concepts and base concepts be available in a variety of media and instructional experiences?
- What experiences and what problems will enable learners to customize and personalize content?

Knowledge Principle #2. Every learner has a zone of proximal development that defines the space that a learner is ready to develop into useful knowledge.

Questions:

- How do we create teaching and learning experiences that support learning within the zone of proximal development for all learners?
- How do we design experiences that lead to the creation of more ZPDs?
- How do we determine/assess what development has occurred?

Knowledge Principle #3. Learners develop useful knowledge in context, in situations similar to those that are meaningful to learners.

Questions:

- What content resources are to be required, recommended and selected/or customized for the student?

The Context or Environment (E) for Learning

In Vygotsky's model, the environment for learning is a fundamental "actor" in the process of learning. It is Vygotsky's belief in the essential role of context and environment in learning that probably resulted in his theory being characterized as a theory of social development. While this is a true characterization to some degree, his theory is more about the role of social interactions in the development of cognition. His theory is an intellectually robust, holistic learning theory, based on context and the interaction of the learner with the environment. In Vygotsky's world, the environment of learning includes the interaction with the faculty mentor, with the learners' peers, with the learning materials and resources, and action and thought within the learner's specific zone of proximal development.

The context of a learning experience for Vygotsky includes all physical aspects and all mental aspects of an environment, the full socio-cultural-temporal environment. When applying this theory to online and distance learning, this means designing for the full range of physical and mental aspects of a learning experience. Vygotsky's theory is a reminder of the importance of designing for the "where, when, with whom and with what resources" of a learning experience.

Principles/questions for the context/environment:

What principles and questions does the life-style and learning style framework suggest be asked about the context and the environment of learning?

Environment/Context Principle #3. Every learning experience has a context or an environment in which the learner works.

Questions:

* Where, when, with whom and with what resources is a learner going to be working?

* Will a learner have choices of resources, choices of learning tools and strategies?

* When will a learner know that he/she knows? What feedback or result from a problem being solved will make the learner's knowledge evident?

Level Six: Assessment

As mentioned in the instructional design principles, assessment is fundamental to the design process. Assessment planning helps to balance the goals of the effectiveness of learning and the efficiency of teaching and learning. However, assessment is complex and challenging, particularly if we acknowledge the goals of enriched student learning. In productive, customized, enriched learning experiences, each learner begins with an existing, distinct base of knowledge representation and enriches that knowledge base in distinct and individual ways. Productive learning experiences mean that all students complete a learning experience with an expanded, yet different, base of knowledge. The goal is that learners learn core principles so that they can be effectively applied, but the particular way the knowledge base is constructed in individuals is unique.

What we can design into instructional planning is that all learners share some of the same experiences and that assessment focuses on the common learning that is achieved. Assessment can also provide for demonstration of knowledge and skills in more complex environments. How this is achieved in online and distance learning requires creativity and flexibility beyond the usual types of assessments.

Given these few caveats, here are some questions that faculty mentors and instructional planners can use as starting points for designing assessment.

Assessment Questions

- How will the learners know the goals and objectives for the learning? It is good to plan for core concepts, practice of core concepts, and customized applications of concepts.

- Will learners be generating a set of their own goals for learning? How will the faculty mentor and learner communicate and agree on goals for learning, particularly for customized applications of concepts?

- Will learners have self-check or peer-check practice sessions on core concepts?

- How often will assessment be done? Will there be weekly expectations?

- In what ways and where will the students be evaluated and graded?

- How will students demonstrate their competency in concept formation? In solving problems?

- If we don't see the students on a regular basis, can we design ways to "see" their minds virtually through conversation and experiences?

This is only a brief beginning on designing and planning for assessment. Here are two resources for following up with more ideas on designing for complex learning. Alverno College in Milwaukee, Wisconsin, has focused on "ability based education" since the 1960s. Their program is based on the concept of "student assessment as learning." It is an interesting approach that has included portfolios, etc., for some time. Descriptions of their programs and their approach to student assessment are at www.alverno.edu/educators/student_as_learn.html. The portfolio projects within the National Learning Infrastructure Initiative (NLII) are also useful in assessment of complex learning. (See http://www.educause.edu/nlii/keythemes/eportfolios.asp.)

Design Questions for Future Planning

The design questions proposed for each of the six design levels focus primarily on the components of the teaching and learning experience. Many external trends and environmental factors are currently impacting the design of instruction, as well, but those are the focus of a future paper.

Summary of Design Levels

The process of the design of instructional planning is gaining respect as the demand for effective and efficient learning grows. The demands are coming as a result of time pressures, budget pressures and a general goal of accountability in our society. We started with a set of design principles that reaffirmed that design work is iterative, that the responsibility for design work is shared jointly among the hierarchical groups of an institution and that instructional planning, when done well, results in delighted and productive learners and faculty/mentors pleased with their roles and their work. These types of outcomes argue persuasively for this new focus on instructional design and planning.

Acknowledgments

With many thanks to Dr. Don Ely, Professor Emeritus at Syracuse University and Dr. Dale Lick of Florida State University and other friends and colleagues for their feedback on earlier drafts of this chapter.

References

Arnone, M. (2002). United States Open U. to close after spending $20-million. *Chronicle of Higher Education*, (February 15), A44.

Bartlett, T. (2003). Take my chair (Please!). *Chronicle of Higher Education*. (March 7), A36-38.

Bates, A. W. T. (1995). *Technology, Open Learning and Distance Education.* New York: Routledge.

Blumenstyk, G. (2001). Temple U. shuts down for-profit distance-education company. *Chronicle of Higher Education,* (July 20), A29.

Boettcher, J. V. (1999, October). Another look at the tower of WWWebble. *Syllabus*, 13(50), 52.

Boettcher, J. V. (2000). How much does it cost to put a course online? It all depends. In M. J. Finkelstein, C. Frances, F. Jewett and B. W. Scholz (Eds.), *Dollars, Distance, and Online Education: The New Economics of College Teaching and Learning* (pp. 172-197). Phoenix, AZ: American Council on Education, Oryx Press.

Boettcher, J. V. and Kumar, V. M. S. (2000, June). The other infrastructure: Distance education's digital plant. *Syllabus*, (13), 14-22.

Boettcher, J. V. and Long, P. (2002). *What's next for teaching and learning leveraging technology?* Paper presented at the meeting of the National Learning Infrastructure Initiative.

Bransford, J. D., Brown, A. L. and Cocking, R. R. (1999). *How People Learn. Brain, Mind, Experience, and School.* Washington, DC: National Academy Press.

Brown, D. G. (2000). Academic planning and technology. In J. V. Boettcher, M. M. Doyle and R. W. Jensen (Eds.), *Technology-Driven Planning: Principles to Practice.* Ann Arbor, MI: Society for College and University Planning.

Bruner, J. S. (1963). *The Process of Education.* New York: Vintage Books.

Daniel, J. S. (1998). *Knowledge media for global universities: Scaling up new technology at the Open University.* Paper presented at the meeting of Seminars on Academic Computing, Snowmass, Colorado, USA.

Dewey, J. (1933). *How We Think* (1998 ed.). Boston, MA: Houghton-Mifflin.

Gagne, R. M. (1965). *The Conditions of Learning.* New York: Holt, Rinehart & Winston.

Green, K. C. (2002). *Campus computing, 2002.* Encino, CA: The Campus Computing Project, www.campuscomputing.net.

Kaufman, R. and Lick, D. D. (2000). Mega-level strategic planning: Beyond conventional wisdom. In J. V. Boettcher, M. M. Doyle and R. W. Jensen (Eds.), *Technology-Driven Planning: Principles to Practice.* Ann Arbor, MI: Society for College and University Planning.

Kolb, J. E., Gabriele, G. A. and Roy, S. (2000). Cycles in curriculum planning. In J. V. Boettcher, M. M. Doyle and R. W. Jensen (Eds.), *Technology-Driven Planning: Principles to Practice.* Ann Arbor, MI: Society for College and University Planning.

Lessig, L. (1999). *Code and Other Laws of Cyberspace.* New York: Basic Books.

Lick, D. and Kaufman, R. (2000). Change creation: The rest of the planning story. In J. V. Boettcher, M. M. Doyle and R. W. Jensen (Eds.), *Technology-Driven Planning: Principles to Practice.* Ann Arbor, MI: Society for College and University Planning.

Pinker, S. (1997). *How the Mind Works.* New York: W.W. Norton.

Riffee, W. H. (2003, February). Putting a faculty face on distance education programs. *Syllabus*, (16), 10-14.

Sherron, G. T. and Boettcher, J. V. (1997). *Distance learning: The shift to interactivity.* CAUSE Professional Paper Series #17. Retrieved March 6, 2003 from the World Wide Web: EDUCAUSE Publications Database.

Vygotsky, L. S. (1962). *Thought and Language.* (E. Hanfmann and G. Vakar, trans.) Cambridge, MA: MIT Press.

Vygotsky, L. S. (1978). *Mind in Society: The Development of Higher Psychological Processes.* Cambridge, MA: Harvard University Press.

Chapter III

E-Moderating in Higher Education

Gilly Salmon
Open University Business School, United Kingdom

Abstract

There are few published reports of structured approaches to developing lecturers for new online roles.   However, both campus and distance learning institutions can offer some experiences in developing lecturing staff to moderate and teach with low cost text-based online conferencing. This role is known as e-moderating. Staff development is often asserted as a key issue in the success of everything from a project, a course or a whole institution to an online environment.   The current climate asserts the importance both for university and college lecturers of adopting a good practice and an understanding of teaching in addition to academic competence.   This chapter considers and explores the knowledge and skills that the best e-moderators have and how they can be recruited, trained and developed.

Introduction and Rationale

The challenge of developing new kinds of online teaching and learning processes, while remaining true to educational or training missions, is at the forefront of the implementation of information and communication technologies in the early 21st century. Alexander, McKenzie et al. (1988) show that staff development is one of the main factors in determining the success of institutional attempts to make the transition to online delivery.

The term moderator has grown up with the use of online text-based discussion and group work, in teaching and learning contexts. In 2000, I first used the term "e-moderating" to capture the wide variety of roles and skills that the online teacher, lecturer or trainer needs to acquire. Supporting learning online through synchronous and asynchronous conferencing (bulletin boards, forums) requires e-moderators to have a wider range of expertise compared to working with face-to-face learning groups. Hence, the role of the lecturer or teacher in higher education needs to change to include e-moderating to match the development and potential of new online environments.

Successful and productive e-moderating is a key feature of positive, scalable and affordable e-learning projects and processes. Regardless of the sophistication of the technology, online learners do *not* wish to do without their human supporters. How many people, for example, have been heard to say, "I'm great at art because of my inspirational computer"? Not any that I've met, on or off-line! Instead, learners talk of challenge and support by their teachers or of contact with the thoughts and the work of others. Most people also mention the fun and companionship of working and learning together. Such benefits do not have to be abandoned if developing online learning results in a cohort of trained e-moderators to support the online learners.

Many words have been written about new technologies and their potential, but not much about what the human supporters of the learning actually *do* online. The greatest impact of all on the quality of the students' learning resides in the way a technology is used and not in the characteristics of the medium itself (Inglis, Ling et al., 2000). Although increasing numbers of learners are working online, few lecturers have themselves learned this way. Therefore, e-moderating is not a set of skills most lecturers have acquired vicariously through observing teachers while they themselves were learning. Many lecturers naturally believe that learning to e-moderate mostly has to do with learning new software or computing skills. This is not the case. In text-based asynchronous environments, a critically important role for the e-moderator is promoting the surfacing and sharing of understanding and knowledge through online writing and dialogue (Barker, 2002). Furthermore, successful e-moderating cannot be achieved by doing what lecturers always did in the classroom. As of yet, there are few online mentors to guide them through step-by-step, nor is there time for long-term apprenticeships. It follows that e-moderators must be specially recruited, trained and developed.

Currently, e-moderating continues to be a labour-intensive service. The UK Open University, for example, works to an average of 25 students per online teaching discussion group. The for-profit University of Phoenix in the US operates in learning groups of eight to 15 students, each with an online teacher. This means that we are likely to witness a growth in demand for online teachers in the next few years.

Definitions and Context

There are many definitions of an online course. These include classroom-based teaching supplemented by lecture notes posted on a web site or by electronic communication such as e-mail. At the other end of the spectrum, materials may be made available and interactions occur exclusively through networked technologies. Currently, in the UK, completely online courses are rare. Most courses are mixed mode or blended in some way.

I use the term online to mean teaching and learning which takes place over a computer network of some kind (e.g., an intranet or the Internet) and in which interaction between people is an important form of support for the learning process. This rules out learning which is purely "resource-based," e.g., learning using some Web-based courseware without recourse to any kind of human interaction. It includes both synchronous and asynchronous forms of interaction and also interaction through text, video, audio and in shared virtual worlds (Goodyear, Salmon et al., 2001). To date, text-based asynchronous computer mediated conferencing or forums have been the most extensively used for teaching and learning in higher education, both on and off campus, and hence I concentrate especially on the roles of e-moderators in asynchronous networked learning environments. Platforms most commonly in use include FirstClass, Blackboard and WebCT, but there are a wide variety of others, including commercial systems and those developed in-house.

E-moderating draws on aspects of both face-to-face teaching and traditional print-based distance teaching. However, it also calls for the introduction of a range of new understandings and techniques that are specific to online delivery. The key factor in e-moderating is that the e-moderator operates for part of the time in the electronic environment along with his or her students or learners.

Online teaching also requires an attitude change for teaching staff. Some researchers argue this is nothing less than a major cultural change (Williams, 2002).

There are two main ways that lecturers in universities and colleges can be developed to engage successfully in online teaching opportunities. One approach is to enhance the technical knowledge of teaching staff to enable them directly to design, develop and produce online materials and teach online with their students. Some of the platforms in use, such as Blackboard, have been developed to make these tasks as simple and as non-technical as possible. However, a second and more common method is to take a specialist approach to staff development. This involves increasing the skills of lecturers to focus on the successful delivery of online teaching and facilitating and the support of online students, usually in combination with their other (off-line) duties. This means that other people also must be recruited, such as instructional designers, graphic designers, computer technicians, multimedia programmers, audio-visual technicians, editors of text, e-librarians and resource providers. (See Inglis, Ling et al., 2000, for further exploration of these issues.)

Competencies and Skills for E-Moderators

Goodyear, Salmon et al. (2001) detail the online teacher's roles as follows:

1. *Process facilitator* — facilitating the range of online activities that are supportive of student learning.

2. *Adviser/counselor* — working on an individual/private basis, offering advice or counseling learners to help them get the most out of their engagement in a course.

3. *Assessor* — concerned with providing grades, feedback, validation of learners' work, etc.

4. *Researcher* — concerned with engagement in production of new knowledge of relevance to the content areas being taught.

5. *Content facilitator* — concerned directly with facilitating the learners' growing understanding of course content.

6. *Technologist* — concerned with making or helping to make technological choices that improve the environment available to learners.

7. *Designer* — concerned with designing worthwhile online learning tasks (both "pre-course" and "in course").

8.   *Manager/administrator* — concerned with issues of learner registration, security, record keeping, etc.

Of these, the most difficult to grasp and achieve are the process roles, e.g., one to five. These are the roles that I call e-moderating.

I have analysed the qualities and characteristics of successful e-moderators — the competencies they should acquire through training and experience (Salmon, 2002) (see Figure 1).

Recruiting Your E-Moderators

Given the required competencies, how do you set about acquiring the right e-moderating staff? Most institutions face re-skilling experienced staff, or adding e-moderation to training programmes for new teaching staff. Currently, it is most unusual for lecturers or tutors to be recruited for e-moderating skills *per se* or for their previous experience in teaching online. In the future, however, I predict that there will be more specialisation, and some categories of staff will demonstrate their particular aptitude for working online.

A long list of relevant teaching qualifications or experiences is unlikely to be found at this stage of the development of online lecturers and trainers. The e-moderators you recruit should, of course, be credible as members of the learning community. I suggest that you try to recruit e-moderators with the qualities from columns one and two of Figure 1, if possible. Teachers who have something of a vision of the importance of online learning in the future and how to prepare themselves to operate successfully and happily within such an environment are those to be spotted and supported (Waeytens, Lens et al., 2002). However, at the moment, there are very few such people available. I tend to select applicants who show empathy and flexibility in working online, plus exhibit a willingness to be trained as e-moderators. Before asking them to work online, you must train them in the competencies described in columns three and four in Table 1. I would expect e-moderators to have developed the skills in columns five and six by the time they had been working online with their participants for about one year.

If lecturers or academics are used to being considered an "expert" in their subject, the levelling effect and informality of online networking can be very challenging for them. As e-moderators, they will probably have to work a little harder to establish their credentials as an experienced professional in the online environment, as opposed to being in a face-to-face group. Even those recruits who are used to developing distance learning materials need to explore how

Figure 1.   Table of E-Moderator Competencies (From Salmon, 2000)

Quality/ Characteristic	RECRUIT		TRAIN		DEVELOP	
	I CONFIDENT	II CONSTRUCTIVE	III DEVELOPMENTAL	IV FACILITATING	V KNOWLEDGE SHARING	VI CREATIVE
Understanding of online process A	Has personal experience as an online learner, flexibility in approaches to teaching & learning. Empathy with the challenges of becoming an online learner.	Is able to build online trust and purpose for others. Understands the potential of online learning and groups.	Has ability to develop and enable others, act as catalyst, foster discussion, summarize, restate, challenge, monitor understanding and misunderstanding, take feedback.	Knows when to control groups, when to let go, how to bring in non-participants, know how to pace discussion and use time on line, understand the five-stage scaffolding process and how to use it.	Can **explore** ideas, develop arguments, promote valuable threads, close off unproductive threads, choose when to archive.	Is able to use a range of approaches from structured activities (e-tivities) to free wheeling discussions, and to evaluate and judge success of these.
Technical skills B	Has operational understanding of software in use reasonable keyboard skills; able to read fairly comfortably on screen, good, regular, mobile access to the Internet.	Is able to appreciate the basic structures of CMC, and the WWW and Internet's potential for learning.	Knows how to use special features of software for e-moderators, e.g., controlling, weaving, archiving. Know how to "scale up" without consuming inordinate amounts of personal time, by using the software productively.	Is able to use special features of software to explore learner's use, e.g., message history.	Creates links between CMC and other features of learning programmes.	Is able to use software facilities to create and manipulate conferences and to generate an online learning environment, able to use alternative software and platforms.
Online communication skills C	Provides courteous and respectful in online (written) communication, able to pace and use time appropriately	Is able to write concise, energizing, personable online messages.	Is able to engage with people online (not the machine or the software), responds to messages appropriately, be appropriately "visible" online, elicit and manage students' expectations.	Is able to interact through e-mail and conferencing and achieve interaction between others, be a role model. Able to gradually increase the number of learners dealt with successfully online, without huge amounts of extra personal time	Is able to value diversity with cultural sensitivity, explore differences and meanings.	Is able to communicate comfortably without visual cues, able to diagnose and solve problems and opportunities online, use humour online, use and work with emotion online, handle conflict constructively.
Content expertise D	Has knowledge and experience to share, willingness to add own contributions.	Is able to encourage sound contributions from others, know of useful online resources for their topic.	Is able to trigger debates by posing intriguing questions.	Carries authority by awarding marks fairly to students for their participation & contributions.	Knows about valuable resources (e.g., on the WWW) and refer participants to them.	Is able to enliven conferences through use of multimedia and electronic resources, able to give creative feedback and build on participants' ideas.
Personal characteristics E	Has determination and motivation to become an e-moderator.	Is able to establish an online identity as e-moderator.	Is able to adapt to new teaching contexts, methods, audiences and roles.	Shows sensitivity to online relationships and communication.	Shows a positive attitude, commitment and enthusiasm for online learning.	Knows how to create and *sustain* a useful, relevant online learning community.

online materials can underpin and extend their teaching. It follows that e-moderators will also need to develop good working relationships with librarians who are rapidly transforming themselves into ICT resource providers. Understanding ICT resource provision is another aspect that can be checked out during recruitment by asking, for example, about a favourite web site for their subject.

It is most important to look at the potential, as well as at the existing skills of recruits. E-moderators will need to know about online communication, rather than only learn the software. They will need to have the ability to provide support and counseling through e-mail, as well as the creativity and flexibility to design and adapt collaborative opportunities for differing purposes, individual and organisational missions and needs. They must be able to work cross-culturally and value diversity. They must be flexible in considering approaches to online assessment, evaluation and achievement. They must understand the benefits of online working and, hence, are able to act as resource guides and monitors. Furthermore, they will have personal metacognitive and adaptable approaches to learning and the ability to reflect and have input into overall course learning processes. The following table suggests some questions to ask to establish such potentials. Skills in a particular platform are unimportant at the recruitment stage. These can be taught and retaught as needed.

To summarise, at the moment there are few people available with the skills I have listed in Figure 1. Most newcomers to e-moderating are more familiar with teaching face-to-face, where they rely perhaps on personal charisma to stimulate and hold their students' interest. So what is most important in recruiting e-moderators? The main enemy of operating successfully as an e-moderator in asynchronous environments is time. Therefore, you need to look for people who will not try to cover everything. But, instead look for those who have student-focused strategies based on encouraging the students to change their view of the world — strategies that are based on what the students do. E-moderator recruits also need to be very organised. There is rarely the same opportunity to improvise as there is in face-to-face teaching. I would consider the most essential skills are empathy online and a flexible approach to working, teaching and learning online. Any recruit will have to be willing to be trained and developed in the e-moderating role. They will need good keyboarding skills and some experience of using computers, including online networking. "Black and white thinkers" are generally to be avoided. What you are recruiting for, to a large extent, is the ability to adapt to technological and online environment changes, as well as to operate successfully using current platforms. Look for people who have cross-cultural sensitivity which includes the ability to handle ambiguity and multiple viewpoints. However, given those requirements, good e-moderators come from many different backgrounds, with very varied learning and teaching experiences. Where they live, their domestic or work commitments or any disabilities that reduce their ability to travel are unimportant.

Training E-Moderators

Even if teachers have an excellent record in conventional settings, it is difficult to predict who will do well in online teaching. Currently, few universities and colleges offer much in the way of training for e-moderating skills and the best methods are yet to be identified (Kearsley, 2000). However, the acquisition of e-moderating skills cannot be achieved vicariously by lecturers observing other online teachers or by looking at exemplary web sites. Enabling lecturers to use technology in their teaching, means providing training that is motivating, attention gaining, relevant and confidence building. A tall order, indeed!

Scaffolding: A Model of Online Development

Figure 2 offers a model of teaching and learning online, researched and developed with business school students and tutors in the Open University over several years and since applied across many learning disciplines, contexts and levels. This model can be used as a scaffold for training and developing e-moderators. Scaffolding suggests a way of structuring online interaction and collaboration, starting with recruitment of interest, establishing and maintaining an orientation towards task-relevant goals, highlighting critical features that might be overlooked, demonstrating how to achieve those goals and helping to control frustration (Wood and Wood, 1996).

It is especially important to concentrate on the communicative aspects of the use of the online learning platform (Monteith and Smith, 2001). Each level of the five-stage model involves somewhat different activities for the participants. What the e-moderator does online, and how much, varies according to the purposes, intentions, plans and hopes for online learning. There is growing evidence that teachers benefit from a developmental approach to learning new techniques (Cornford, 2002).

There are certain key stages in the progression of the trainee e-moderator. First, there is a crucial understanding that gradually increasing the comfort of online learners will increase participation and completion rates. Second, an appreciation develops that the design of online activities and interaction is as important as sophisticated, but non-dynamic, design and delivery of content. Third, that the evolving role of the e-moderator, who is much more than just a facilitator or responder to questions, will make or break the experience for the learners. Fourth, there is a recognition that there is considerable evidence that people become more independent and more responsible for their own development as they move through the model, whether in structured or informal learning settings. To date, most lecturers have acquired e-moderating skills through informal networking and self-teaching, often in a situation of severe time poverty (Bennett, 2002).

I will now outline the model, before going into detail. Facilitating individual access is an essential prerequisite for conference participation (stage one, at the base of the flight of steps). Stage two involves individual participants establishing their online identities and then finding others with whom to interact. At stage three, participants give and receive information freely to each other. Up to and including stage three, a form of co-operation occurs, i.e., support by the online group for each person's goals. At stage four, course-related group discussions occur and the interaction becomes more collaborative. The communication depends on the establishment of common understandings. At stage five, participants look for more benefits from the system to help them achieve personal goals, explore how to integrate online learning into other forms of learning and reflect on the learning processes.

Figure 2.   (From Salmon, 2000)

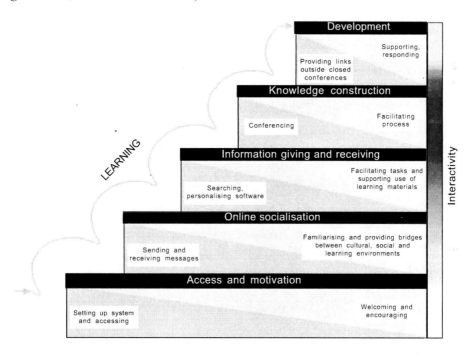

Each stage requires participants to master certain technical skills (shown in the bottom left of each step). Each stage calls for different e-moderating skills (shown on the right top of each step). The "interactivity bar" running along the right of the flight of steps suggests the intensity of interactivity that you can expect between the participants at each stage. At first, at stage one, they interact only with one or two others. After stage two, the numbers of others with whom they interact, as well as the frequency, gradually increases, although stage five often results in a return to more individual pursuits.

Given appropriate technical support, an e-moderator, and a purpose for taking part in online networking, nearly all participants will progress through these stages of use in online learning. There will, however, be very different responses in terms how much time they need at each stage before progressing. The model applies to all online learning platforms. If experienced participants are introduced to software that is new to them, they will tend to linger for a while at stages one or two, but then move on quite rapidly up the steps again.

The chief benefit of using the model to design development processes for e-moderators is that there is a greater readiness by e-moderators, once trained, to contribute to student learning online. E-moderators who understand the model and apply it should enjoy online learning and find that their work runs smoothly. They will spend much less time and achieve a more productive experience for their students. If suitable technical and e-moderating help is given to participants at each stage of the model, they are more likely to move up through the stages, to arrive comfortably and happily at stages three through five. These stages are the ones that are most productive and constructive for learning development purposes.

Stage One: Access and Motivation

For e-moderators, being able to gain access quickly and easily to the online learning system is one key issue at stage one. The other is being motivated to spend time and effort.

The participant needs information and technical support to get online and strong motivation and encouragement to put in the necessary time and effort. Like learning any new piece of software, mastering the system seems fairly daunting to start with. However, it is important to reassure lecturers that successful achievement of e-moderation does not depend on previous computer literacy (online networking often appeals to inexperienced computer users). At this stage, computer skills will vary enormously. Use of e-mail is almost universal, whereas competent, effective communication with it is not. And few people are really skilled in using asynchronous groupware. Most people will be unfamiliar with the tools you choose to use. However, many trainee e-moderators need some form of individual technical help at this stage, as well as general encouragement. Problems are often specific to a particular configuration of hardware, software and network access or else related to the loss of a password. Access to technical support needs to be available, probably through a telephone helpline, particularly when the trainee is struggling to get online on his or her own. However, it is a mistake to offer extensive face-to-face workshops to lecturers in an attempt to enable them to feel more comfortable with the technology. They need to develop these skills in the relevant context of their own teaching. The simplest approach is to emulate the student experience for the lecturers, since few of them have been online learners.

Another aspect of access is special needs of various kinds. In the spirit of wide diversity and empowerment, it is good that the disabilities of users with special needs are not usually obvious online. It is normally impossible to tell from the messages in a conference that a participant or an e-moderator has restricted vision or hearing or problems of mobility, unless that person wishes to volunteer the information. People who have problems with their speech or hearing are not at a disadvantage in text-based conferencing. However, those who have problems with their vision or physical movement may well find that the keyboard and screen prevent them from doing as much as they would like. Dyslexics still have some difficulties online, even with electronic help available. It is very important that trainee e-moderators understand the nature of special needs and facilitate learning through the medium.

Strong motivation is a prime factor at this first stage, when participants have to tackle the technical problems. Nearly all studies have shown that trainee e-moderators are extremely resistant to insistence on their participation (Salmon, 2000; Collis and Moonen, 2001; Inglis, Ling et al., 2001), so carrots are much more likely to be successful than sticks. No doubt, all staff will have heard horror stories about the time required for e-moderation processes. So they need reassurance, training and clear agreements about the time required from them.

Stage Two: Online Socialisation

In stage two, participants get used to being in the new online environment. Many of the benefits of online learning in education and training flow from building an online community of people who work together at common tasks.

Stage two is a critical stage and cannot be left out. It provides the motivation and creates the important building blocks of professional development. Networked learning offers the "affordance" of online socialising. Affordance means that the technology enables or creates the opportunity (Gaver, 1996). In the case of online conferencing, it has an inherent social component, as well as the ability to convey feelings and build relationships. However, online learning will not in itself create social interaction. Sensitive and appropriate online conference design and the intervening support of more experienced e-moderators will cause the socialisation to occur. It is essential to offer trainee e-moderators the experience of working online *with others*.

A sense of continuity, a connectedness with time and place, and a connectedness with others contribute to online socialising. Our own internalised set of instructions for how to behave and how to make judgements, for feeling comfortable together and "at home in one's world," and the reassurance of the familiar — these all help enable us to find our roots in the social world. When online teaching fragments and expands this sense of time and place, the usual pillars of well-being may be less available. There is evidence at stage two that all trainee e-moderators struggle to find their sense of time and place in the online environment. Hence, the importance of enabling induction into online learning to take place with support and in an explicitly targeted way.

Stage two participants recognise the need to identify with each other, to develop a sense of direction online and they need some guide to judgement and behaviour. A wide range of responses occurs. Some are initially reluctant to commit themselves fully to public participation in conferencing, and should be encouraged to read and enjoy other's contributions to the conferences for a short while, before taking the plunge and posting their own messages. This behaviour is sometimes known as "lurking," although the term can cause offence! "Browsing" or "vicarious learning" are perhaps safer words. Browsing appears to be a natural and normal part of socialisation into online learning and should, therefore, be encouraged for a while as a first step. When participants feel at home with the online culture, and reasonably comfortable with the technology, they move on to contributing. E-moderators really do have to learn to use their skills to ensure that participants develop a sense of community *in the medium*.

O'Brien tells us, from nursing education online, that integration of beliefs and practices of individuals from all social and cultural groups in society is very important for professional development (1998). This is applicable for all online groups, especially as we teach and learn in increasingly international communities. The idea is simply that we should encourage participants to identify and share their own beliefs and values, and acknowledge that they are different from those of others.

Stage Three: Information Exchange

A key characteristic of online learning is that the system provides all participants with access to information in the same way. At stage three, participants start to appreciate the broad range of information available online. Information exchanges flow very freely since the "cost" of responding to a request for information is quite low. In my experience, participants become excited, even joyful, about the immediate access and fast information exchange. They also show consternation at the volume of information suddenly becoming available.

E-moderators can be helped to become independent, confident, and enthusiastic about working online at this stage. Trainee e-moderators need to know how to exchange information in online conferences and forums. Information exchange proficiency is essential before they move on to full-scale interaction in stage four. Demands for help can be considerable because their experience in searching and selection may still be low. There can be many queries about where to find one thing or another online.

In summary, trainee e-moderators develop a variety of strategies to deal with the potential information overload at this stage. Some do not try to read all messages. Some remove themselves from conferences of little or no interest to them, and save or download others. Others try to read everything and spend considerable time happily online, responding where appropriate. Yet, others try to read everything, but rarely respond. These participants sometimes become irritated and frustrated. They may even disappear offline.

Stage Four: Knowledge Construction

Familiarity with the technology must be achieved by this stage. If familiarity is not achieved, then it will only provide a distraction from the much more demanding experience of learning and development in unknown territory and the relationships that now develop. At this stage, trainee e-moderators will become very interested in like-minded communities that are available to them online, especially those from their own areas of interest or disciplines.

The issues that can be dealt with best at this stage are those that have no one right or obvious answers, or ones that participants need to make sense of, or a series of ideas or challenges. These issues are likely to be strategic, problem, or practice-based ones. Most importantly, the development of tacit knowledge and its impact on practice can be very strong at this stage, and especially important in practice-based or clinical-based learning, such as management or medicine.

Developing advanced skills in e-moderating is important at this stage. The best e-moderators demonstrate online the highest levels of tutoring skills related to building and sustaining groups. Feenberg (1989) coined the term "weaving" to describe the flow of discussion and how it can be pulled together. Online learning makes weaving easier to promote than learning in face-to-face groups, since everything that has been "said" is available in the conference text. The best e-moderators undertake the "weaving." They pull together the participants' contributions by, for example, collecting statements and relating them to concepts and theories from the course. They enable the development of ideas through discussion and collaboration. They summarise from time to time, span wide-ranging views and provide new topics when discussions founder. They stimulate fresh strands of thought, introduce new themes and suggest alternative approaches. In doing all this work, their techniques for sharing good practice and for facilitating the processes become critical.

Trainee e-moderators need to learn how to add value to the online discussion forum or conference contributions at level four. First, the contributor needs to be acknowledged in order to be "heard." Secondly, online contributions will be recorded, made available for others to read and become a form of inventory. The e-moderator's role is that of a recorder creating the inventory to be surfaced and used by others. In a collective conference, personal "reflections" may be responded to in various ways. One person may need more time to explore issues, while another may reach conclusions quickly and then become impatient with those who are still thinking. It is important that the e-moderator avoids the temptation to discount the experience in any way or to counter it and enter into argument. Instead, he or she can draw on the evidence that is presented to try and explore overall conclusions. Thirdly, the e-moderator should comment at an appropriate moment on the sufficiency of the data being presented and, fourthly, on the quality of the argument around it. These ways ensure that the experiences, while valued, are not necessarily considered complete in themselves. The e-moderator is thereby modelling ways of exploring and developing arguments.

The locus of power in more formal learning relationships is very much with the tutor, lecturer or academic expert. In online learning at stage four, however, there is much less of a hierarchy and greater potential for individual responsibility for development.

Stage Five: Development

In higher education, e-moderators need to be able to engage in reflective practice themselves (Orsini-Jones and Davidson, 1999) and to be very democratic and open about their roles (Hunt, 1998). The challenge is to enable participants to recognise the narrowness of their own experience and be open to other evidence. The e-moderator should learn to prompt, encourage and enable such openness, while acknowledging personal experience. Sensitivity and courage may be needed to explore an experience with well-established, well-focused people!

At stage five, trainee e-moderators will start to display considerable confidence, independence and autonomy in the online environment. Indeed, they frequently start reverting to levels of confidence that they display in their more familiar offline worlds! They will become responsible for their own teaching and development through computer-mediated opportunities and need little support beyond that already available. Rather different skills come into play at this stage. These are skills of critical thinking and the ability to challenge the "givens." Meta-cognitive learning skills are focussed on the impact of their teaching in new contexts or how they might apply concepts and ideas.

When trainee e-moderators are learning through a new medium, such as online learning, their understanding of the processes of using the software and of the experience of teaching in new ways is being constructed, too. It is, therefore, common at stage five for trainee e-moderators to reflect on and discuss how they are networking and to evaluate the technology and its impact on the learning processes for their arenas of teaching.

Successful E-Moderator Training

E-Moderating Training Design

Columns three and four in Figure 1 provide competencies and objectives for training e-moderators. The model described in Figure 3 can be used to provide a framework for training. The training programme should be intrinsically motivating and lead to competent practice. The task is, therefore, to develop a programme that, while providing the development of essential basic skills, such as online confidence and competence, also represents as closely as possible the realities of teaching and learning online. Providing training of this kind is the best way of scaling up from the innovators and early adopters to the bulk of the lecturing and teaching team (Carlson and Repman, 2002).

In order to indicate to trainee e-moderators that learning to teach online needs to be undertaken online, the training programme should use networked technologies and be accessed from the trainees' own machines. Furthermore, the training should focus on pedagogical knowledge, built up through personal and collective reflection on practice, rather than on merely acquiring a technical grasp of the hardware and software. Most importantly, trainee e-moderators must have the experience of working, participating and themselves learning through others online for the training to be successful (Tsui and Ki, 2002). Such an approach sends a powerful message and provides an invaluable confidence builder.

The programme should create a series of "micro-worlds" in which the trainees can interact with each other, with the e-convenors (trainer of the trainers, e-moderator of the e-moderators) and with the software, before progressing to the next stage. They should be advised of appropriate ways of undertaking the tasks, but could also construct their own approach. They need to learn to use the software as a matter of routine, while raising their awareness of the teaching and learning aspects. The importance ascribed in constructivism to the building of relationships between new and existing knowledge (Bruner, 1986) means the careful choice of titles for conferences, and the use of familiar metaphors for explaining aspects of online learning.

E-moderator training must be meaningful and worthwhile for all lecturers if it is to be judged a success. Each stage in the training model should provide a "scaffold" or guide for training teachers (both new and experienced) from novice to expert status in and through the online environment itself. Basic computer literacy can be assumed. If teachers lack basic computing knowledge, they should be offered familiarity training and experience first. However, even advanced skills, for example, in using the Internet, will be insufficient. Experienced online users should be reassured that the programme is about e-moderating skills and not general technological competence. The process should be interactive (with other trainees and online trainers), with additional downloadable support material.

The programme should include training in declarative knowledge (What is this icon?) and procedural knowledge (e.g., How do I send a message?). But, it should focus mainly on more strategic knowledge (What can I do with my e-moderating skills?). However, trainees should acquire these various kinds of knowledge in an integrated way. The online training programme should not only be about acquiring new skills, but it should also help trainees to explore their attitudes about online learning and its meaning for their own teaching.

Helping trainees to control their frustration is a key aspect of learning to learn online. A balance between a trainee struggling with too much complexity and being given enough involvement in the task needs to be achieved. Give more help when trainees get into difficulties and less as they gain proficiency.

I suggest evaluation should be based on tracking the trainees through the stages in the programme by a series of online conferences and questionnaires of a quantitative and qualitative nature. A certificate of completion should be provided, as well as other motivators and incentives, if possible.

Action Based E-Moderator Training — A Model Design

Offering a small piece of information, which I call the "spark," and then asking for some action or reaction on the part of trainees has proved the cheapest, easiest and most effective design to date. This is called "e-tivities" (Salmon, 2002). It has the advantage that little specialised technical knowledge is needed.

It is also cheap and easy to set up within an existing online platform in use in a university or college.

Each session is offered for one week. The programme lasts for five weeks. Participants are expected to take part for between three to five hours a week. It works best if they log on for a half an hour each day, but many other patterns of participation are also possible.

Session 1 includes reading résumés, writing your own résumé, practising writing messages, practising sending e-moderating messages, exploring online chat facilities, encouraging contributions, using help, learning to quiz, evaluating contributions, and beginning the practice of reflecting.

Session 2 includes getting the feel of online working, sharing experiences of being online, responding encouragingly, writing encouraging replies, identifying good and bad practice, practising expressing emotion, determining how frequently you should be online, exploring the nature of vicarious learners online, establishing your group, and more reflecting.

Session 3 e-tivities include evaluating messages for encouragement, weaving, summarising, compiling information, being encouraging without answering questions, information exchanging, and building on reflection with others.

Session 4 e-tivities include creating e-tivities, setting objectives, getting everyone up to the same level, asking questions, acknowledging, identifying new knowledge, producing a summary, becoming more knowledgeable, working in different size groups, and applying reflection to practice.

Session 5 e-tivities include exploring meanings, identifying ways of helping adults learn through online interaction, planning for personal development in e-moderating skills, planning for further development in your e-moderating skills, building your development plan for further improvement in your e-moderating skills, identifying examples of interventions, building resources to aid development of your e-moderating work in future, reflecting on overall experience, and farewells.

Exemplar: University of Glamorgan E-College Project: Online Staff Development in Blackboard

The core activity of the University of Glamorgan (http://www.glam.ac.uk) is the traditional delivery of courses at the University campus, as well as through agreements with its Associate and other Partner Colleges. Most lecturing staff have experience only in face-to-face methods of teaching. A few have been involved with traditional print-based, distance learning. It was decided to build on this experience and success and to work with partners in the public, private and voluntary sectors to widen the accessibility of the University's Business and

Management Courses through new methods of delivery. This work is intended to pilot the University through its Enterprise College (e-college) initiative (http://www.enterprisecollegewales.co.uk).

The concept of the project is based upon forming an alliance of complementary organisations in the commercial, educational, media, communications, public and voluntary sectors to deliver training and skills development. The e-college initiative provides the additional flexibility of training and support of the University's students through online entrepreneurial programmes. The flexibility of online delivery removes barriers and reaches an increased constituency of individuals, businesses (particularly those in the SME sector), and public and voluntary sector bodies. European Structural Funds support the development of course material and the delivery of the training for staff. Students are able to study at home, at work and on the college campus. They also have access to leased computer equipment installed in their homes. This initiative provides a significant opportunity for the University to evaluate the development, delivery and assessment of e-learning and to create a pedagogy or androgogy for this form of education.

In order to provide staff with some expertise in e-moderating, the e-college undertook a major staff development programme. Unusually, staff development was put in place prior to going live with students and staff appreciated this development.

The programme was based on an online asynchronous participative programme in Blackboard. It used the five-stage model of e-moderating development as a framework and interactive activities (e-tivities) to maintain interest and interaction between participants (see Figure 1). The staff development programme was built and run on the e-college's server. Staffs were expected to take part around three hours per week over a five-week period in September and October 2001. Overall aims of the training programme were:

- Provide lecturers with the skills to access and use Blackboard conferencing and to undertake a range of tasks online.

- Provide lecturers with the experience and confidence to use the online discussion system as a key resource in building a student-based online learning community, and enable mobilisation of the learning of those students through simple interactive and e-moderated participation (called e-tivities).

- Enable lecturers to become active members of an online community for E-college e-moderators participating in and contributing to the College's successes, achievement and online interaction.

Evaluation

The online programme introduced a set of simple motivational goals, by requiring participants to reflect "deliberately" on learning at each stage. They were encouraged to take part, to post at least one message at each of the five levels, to contribute to the "reflections" conferences, to complete their exit-questionnaires — and only then to ask for their certificate of completion.

Results

Thirty-four lecturers involved in the E-college project started the course on or around September 10, 2001. Twenty-seven successfully completed the course by mid-October 2001.   Another seven continued to work through the online activities more slowly.

Although staff found the online e-moderation development programme very challenging, nearly all appreciated the opportunity to take part, and felt that they had achieved the objectives. Some of their final reflections are copied below:

"Some interesting reflections posted ... it is heartening to know that others are feeling the same as me — being an e-learner has made me see the other side of the coin. I have learned a lot about learning online — a whole range of emotions from feeling very lonely at times, experiencing happiness when I achieve something and guilt when I know I am not contributing as much as I should be." CH October 01

"It also made me realise that e-moderating is not something you can do in small parcels of time (the odd hour between classes). It needs more attention and thought than that." ES December 01

"I can't believe I've actually reached this stage — final reflections. Overall, I have found the course beneficial. I feel that my confidence in online learning has increased and my navigation skills have certainly improved (although they still are far from perfect). I also feel that I have a better understanding and knowledge of e-learning. This course has, however, also made me realise just how much more I've got to learn — I think there is a very steep learning curve ahead of me." GG October 01

"I have enjoyed the course and I feel I have learned a lot. I now feel more confident about my navigation skills and a little more confident about my

e-moderating skills. I will be even more confident once I have tried it for real. Unlike some, I enjoyed the early stages. Finding out so much about each other helped me feel even more the importance of online socialisation."
NJ October 01

Dr. Norah Jones, the Project Manager, wrote:

"The programme was a success at two levels. First, it enabled us to gain greater confidence in using the software package (Blackboard) and secondly we were able to appreciate the need to fully engage e-learners. We also benefited by staff from many different locations in Wales working together online and getting to know each other through the development programme. We see this form of staff development essential and plan to spread development of this kind across the University for other e-learning projects."

Developing E-Moderating in Service

After the training and when e-moderators start to work with their online learners, development needs to continue. During this phase I'd expect emphasis to be put on an understanding of the benefits and role of creative thinking in e-moderating and in the development of online group activities (see columns five and six in Figure 1). As experience is gained, e-moderators also start to display considerable confidence and skill in keeping their online teaching fresh, alive and varied by the use of creative thinking and inspired activities, typically going beyond the use of online forums only for discussion.

There are two main ways that development can occur. One is through online networking with others, and the other is through peer to peer visits to observe online processes.

Teacher education offers an example of building online learning communities with an impact on professional practice, going further than what is possible in specific training events (Leach and Moon, 1999; Selinger and Pearson, 1999). By working in such a community, participants can extend their networking beyond the institution in which they work. They can also work with others from different educational traditions. Selinger shows us that this aids their attempts to seek out and understand new ideas and opinions. Thus, teacher trainees explore new ways of tackling everyday problems and report the results to the online community.

To gradually build up appropriate and consistent e-moderating practice and ensure quality, you need to set up monitoring of your e-moderators' work. You may, like the Open University, wish to base this on a peer review system. It is better to review and monitor the work of e-moderators *online*. I suggest you make sure that the reviewers are fully comfortable and competent themselves as e-moderators, so they don't apply old paradigms of teaching and learning to the new environment! I suggest monitoring that concentrates on the key issues in Section 9. If e-moderators are coping with these issues, then you can be sure that not only are their skills building up, but participant satisfaction also will be growing. Of course, another important way of determining the success of the work of the e-moderators is to explore the responses of the participants, themselves.

E-Moderating Costs

Some vendors of online learning solutions make their return-on-investment cases by disposing entirely of lecturer or trainer costs. Disposing of the e-moderator is very rarely appropriate in higher and further education. Enormous value is added to the student experience by skilled e-moderation. Increases to the student-to-lecturer ratio are often considered detrimental to student learning and major causes of lecturer pressures. However, skilled and trained e-moderators can often handle large numbers of students online. The student-to-e-moderator ratio depends largely on the purpose of the online learning.

Costing each activity related to online working is difficult, but not impossible. Much depends on the assumptions behind the figures. For example, I compared the estimated costs of training Open University Business School e-moderators face-to-face with the actual costs of training them online. My estimates were based on costs in 1996 of a face-to-face weekend for 180 e-moderators, drawn from all over the UK and Western Europe, including travel and subsistence, attendance fees and set up costs, excluding staffing costs and overheads. These came to roughly £35,000 in 1996. The actual costs of the *online* training for 147 e-moderators totalled roughly £9,000, again without including staffing costs and overheads. The two sets of figures do hide quite a few assumptions, but the cost advantage was apparently considerable in that particular context. For the business school, there is a very substantial competitive advantage in having a large cohort of trained e-moderators for all courses that include online learning now or will do so in the next few years. This advantage, if it could be costed, is probably worth far more than the total cost to date of developing Internet technologies in the school!

Here are some ways of keeping e-moderating costs down:

Keeping E-Moderating Costs Down

1. Make clear decisions about roles and numbers of e-moderators that you will need and the student-to-e-moderator ratio.

2. Keep your e-moderator support to students focused and specify what you expect them to do and when — if necessary, publish total number of hours per week or month available to participants.

3. Establish early on how much e-moderators should be expected to do and what reasonable expectations there are on the part of students.

4. Ensure they are trained in advance of starting work with their students.

5. Train e-moderators online, rather than face-to-face.

6. Train e-moderators using the online platform, itself, thus creating confidence in the platform, as well as creating an e-moderating skill base.

7. Ensure that e-moderators can upload and download messages offline if they wish.

8. Train them how to use your conferencing software or platform software to best their advantage to save time.

9. Look into transfer of costs of hardware, software and connection to students, perhaps with grants for those unable to afford the cost.

10. Set up good helpdesk and online support systems, and encourage competent students to support others, leaving more of your e-moderators' online time for learning related e-moderating skills.

11. Use existing resources and knowledge constructed online as much as possible, rather than develop materials and/or pay for expensive third party materials.

12. Develop systems for re-use of online conferencing materials.

13. Build up economies of scale as rapidly as possible — choose only systems that can be expanded cheaply.

Conclusion

In empowering teachers and academic staff to become e-moderators, we need to deploy an underlying approach to transformative learning for their development involving a process of deconstruction and reconstruction (Vieira, 2002).

Dominant paradigms do need to be challenged (Rolfe, 2002). We need to get below the surface and beyond the myths already grown up around the nature of e-moderating and working online and get into offering real development help.

Currently, the "richness" of the Web depends largely on its volume and the multimedia presentation of *information*. However, I believe the future brings us greater *interaction* — and interaction is fundamental to learning, so long as it is appropriately e-moderated and embedded in the overall learning methods. From these small beginnings, a new body of knowledge and practice will build up for e-moderators that will transfer again and again, even as more connected technologies become available. The need for skillful e-moderation will not disappear, regardless of how sophisticated and fast-moving the technological environments become. E-moderators add the *real* value! I think that the *most* successful teaching and learning organisations and associations will be those that understand, recruit, train, support and give free creative rein to their e-moderators, while addressing the natural fears of loss of power and perceived quality from traditional teaching staff.

References

Alexander, S. et al. (1988). An evaluation of information technology projects for university learning. *Committee for University Teaching and Staff Development*. Canberra: Australian Government Publishing Service.

Barker, P. (2002). On being an online tutor. *IETI*, 39(1), 3-13.

Bennett, R. (2002). Lecturers' attitudes towards new teaching methods. *The International Journal of Management Education*, 43-58.

Bruner, J. (1986). *The Language of Education. Actual Minds, Possible Worlds*. Cambridge, MA: Harvard University Press.

Carlson, R. D. and Repman, J. (2002, July/September). Show me the money? Rewards and incentives for e-learning. *International Journal on E-Learning*, 9-11.

Collis, B. and Moonen, J. (2001). *Flexible Learning in a Digital World: Experiences and Expectations*. Falmer, London: Routledge.

Cornford, I. R. (2002). Reflective teaching: Empirical research findings and some implications for teacher education. *Journal of Vocational education and Training*, 54(2), 219-235.

Feenberg, A. (1989). The written word. In R. D. Mason and A. R. Kaye (Eds.), *Mindweave, Communication, Computers & Distance Education*. Oxford: Pergamon.

Gaver, W. (1996). Affordances for interaction: The social is material for design. *Ecological Psychology*, 8(2), 111-129.

Goodyear, P. et al. (2001). Competencies for online teaching. *Education Training & Development,* 49(1), 65-72.

Hunt, C. (1998). Learning from learner: Reflections on facilitating reflective practice. *Journal of Further and Higher Education,* 22(1), 25-31.

Inglis, A. et al. (2000). *Delivering Digitally.* London & Sterling: Kogan.

Kearsley, G. (2000). *Online Education: Learning and Teaching in Cyberspace.* Belmot, CA: Wadsworth/Thomson Learning.

Leach, J. and Moon, B. (1999). *Recreating Pedagogy. Learners and Pedagogy.* London: Paul Chapman.

Monteith, M. and Smith, J. (2001). Learning in a virtual campus: The pedagogical implications of students' experiences. *IETI,* 38(2), 119-127.

O'Brien, B. S. (1998). Opening minds: Values clarification via electronic meetings. *Computers in Nursing,* 16(5), 266-271.

Orsini-Jones, M. and Davidson, A. (1999, July). From reflective learners to reflective lecturers via WebCT. *Active Learning,* 10, 32-38.

Rolfe, G. (2002). Reflective practice: Where now? *Nurse Education in Practice*, 2, 21-29.

Salmon, G. (2000). *E-moderating: The Key to Teaching and Learning Online.* Falmer, London: Routledge.

Salmon, G. (2002). *E-tivities: The Key to Active Online Learning.* Falmer, London: Routledge.

Selinger, M. and Pearson, J. (eds.). (1999). *Telematics in Education: Trends and Issues.* Oxford: Elsevier.

Tsui, A. B. M. & Ki, W. W. (2002). Teacher participation in computer conferencing: Socio-psychological dimensions. *Journal of Information Technology for Teacher Education,* 1, 23-44.

Vieira, F. (2002). Pedagogic quality at university: What teachers and students think. *Quality in Higher Education,* 3, 256-272.

Waeytens, K. et al. (2002). Learning to learn: Teachers' conception of their supporting role. *Learning and Instruction,* 12, 305-322.

Williams, C. (2002). Learning on-line: A review of recent literature in a rapidly expanding field. *Journal of further and Higher Education*, 26(3), 263-272.

Wood, D. and Wood, H. (1996). Vygotsky, Tutoring & Learning. *Oxford Review of Education*, 22(1) 5-15.

Chapter IV

Community-Based Distributed Learning in a Globalized World

Elizabeth Wellburn
Royal Roads University, Canada

Gregory Claeys
Aboriginal Development Consultants, Canada

Abstract

Through information technologies, there is an increasing connectedness of people in both economic and educational domains. The globalized educational environment is seen by some to be an answer to poverty and other problems through enhanced distributed learning opportunities, while others are concerned that globalization of education is leading to homogeneity and a loss of autonomy for cultures and communities. The authors of this chapter maintain the viewpoint that respectful partnerships can be developed where communities control how their knowledge is used and retained, while at the same time tapping into the potential that the new technologies have to offer and maintaining an appropriate level of quality. Two pilot programs for Indigenous learners, built with this philosophy in mind, are described.

Introduction

For the main part, this short chapter is a relatively personal account of a view toward a consultative, community-based process of distributed learning in the global context and has as its heart the experiences of being involved in two distributed learning pilot projects with Indigenous communities in western Canada. As a sub-theme, this chapter also incorporates a broad-brushed picture derived from current literature on globalization's impact on educational issues.

It is no coincidence that topics for debate surrounding the general economic concept of globalization are also relevant to a discussion of the creation and implementation of distributed learning. The same technologies that have led to globalization in the production of goods and provision of services (allowing multinational corporations, call centres, etc.) also allow learning at a distance in ways that have never been possible before and, as is the case with all educational innovation, there are many complex factors to consider.

The Globalized World

In the most general sense, globalization can be defined as an increasingly connected economic space (Stromquist and Monkman, 2000). A main argument in support of globalization is related to the economies of scale that can be achieved through shared knowledge and to the lofty notion that world peace could be possible through the interconnectedness of economies if global sharing and interdependence were to take place. This view maintains that any nation, state or other grouping of individuals would have a huge incentive to cooperate peacefully if all other nations, states or groups were valued economic partners. Mussa (2000) describes this by referencing historical and current relationships of trade, social attitudes and technologies in a paper prepared for the International Monetary Fund titled *Factors Driving Global Economic Integration.* A statement from this paper is as follows: "Unwelcome efforts to exert control over an alien people, especially in the face of armed opposition, tend to be very expensive in blood and treasure. In contrast, devoting resources to domestic economic development through efficient investments in physical and human capital and development and exploitation of new technologies is an attractive and reliable path to improved national economic well-being" (The End of Empire section, paragraph 4). Another typical reference supporting globalization is found in a publication by the World Bank (2000), which states: "There is compelling evidence that globalization has played an important catalytic role in accelerating growth and reducing poverty in developing countries … in a globalized world, societies everywhere gain from poverty reduction" (p. 1). And similar views are expressed by the Committee for Economic Development (CED, 2000) and many others. As 20th century economic development programs and policies have looked to education as a tool to reduce poverty and as that century ended, links between globalization, greater access to education, and increased standards of living were frequently being drawn. "A greater amount of educational attainment indicates more skilled and more productive workers who, in turn, increase an economy's output of goods and services. In addition, the level and distribution of educational attainment has a strong impact on social outcomes, such as child mortality, fertility, education of children, and income distribution" (Barro and Lee, 2000. p. 1).

In contrast to this are widely voiced, anti-globalization concerns. Opponents of globalization point out that to achieve the economies described by proponents of globalization, the gap between rich and poor is widened, the quality of what is produced is lowered, and the environment is placed at risk. As well, there are concerns that human diversity and cultural integrity are being destroyed or watered-down through a highly competitive, globalizing process that eliminates all but the cheapest and most mainstream options. In short, those who oppose globalization are expressing their dismay that it leads to the westernizing or "coca-cola-ing" of the planet and that globalization ultimately only truly benefits a very small, elite group. The most negative view of globalization is that it is quite simply a corporatized continuation of the colonization that has already caused untold destruction to Indigenous communities[1] (e.g., Battiste, 2000; Stromquist, 2002). In this view, globalization represents unstoppable "rightward" thinking "in its rejection of any government policies that have even the slightest tendency to redistribute wealth or status" (Lowi, 2002, p. 18). Shades of this viewpoint are sometimes reflected in terminology referring to globalized education (especially when packaged with technology), such as "anorexic education" (Blackmore, 2000), or "education as a commodity." Egea-Kuehne (2003) describes a "mercantile approach to education … currently being used in the global marketing of knowledge over Internet and [falling] into what has been called the 'industrial metaphor.'" And Tabb (2001) states that online programs, "owned and controlled by management … take knowledge from the heads and hearts of teachers and put it into CDs and online courses, creating an interchangeable education that can be as standardized as Starbucks or Wal-Mart."

Falling into neither of the two extremes described above, the vision presented in this chapter is that globalization is not going to go away and that it has enormous potential to be both positive and negative with respect to how individuals and communities will ultimately conduct their lives. Working with a philosophy of maximizing the positives through reflection and effort, educators and educational decision-makers can incorporate technology and work within the concept of globalization as the context to nurture skills and build opportunities at a community level. In doing so, and especially by being creative and innovative in the use of technologies and respecting the rights of communities to retain their knowledge and maintain control over decisions on how it is used, partnerships can be formed in which customized education and training are designed to meet cultural needs. This implies a conscious avoidance of the "one-size-fits-all" vision and a movement towards what Dighe (2000) refers to as the "discourse on diversity."

The authors of this chapter believe that this discourse has deep roots and must include a consideration of how individualistic versus community-focused models of the delivery of education ultimately have an impact on the move forward in global interdependency, along with the maintenance of community identities. This world view difference (parallel to the contrast between a linear, behavioristic view of learning and the diversity of learning paths inherent in cognitive constructivism[2]) becomes apparent when, as individuals, we disregard the community perspective and solely focus on our individual needs. For example, if mass marketing of "quality controlled" training becomes a priority, then an appreciation of the unique needs of communities can be lost. Although mass marketed learning experiences can facilitate the acquisition of necessary skills in a training situation, the social goals of education may suffer if local community input (with respect to both content and delivery of education) is not incorporated. In our view, the best strategies are conscious of the differences between education and training and, as much as possible, incorporate constructivistic concepts from both a cognitive community perspective and an individual perspective. In this way balance is provided to begin to address all of the other requirements of community connectivity and cross-cultural or global connectivity and interdependency.

Globalization and Social and Educational Change

Globalization is a controversial concept that clearly relates to technological advancement, which, in turn, has an impact on how education is conceived. Stromquist and Monkman (2000) state that globalization processes (including the skills requirements of technological advancement) have created a climate where "the influence of business and its accompanying values and norms are spreading throughout the world," and ask, "What consequences will this ultimately have for education?" (p. 19). In much of the literature, globalization is seen to have created a competitive "have" vs. "have-not" atmosphere with respect to the acquisition of technology-based skills for citizens (e.g., see Dighe, 2000; Carnoy,

2000). Acquiring knowledge workers who are fluent in technology has become a widely accepted goal of many corporations and nations wanting to be part of what is increasingly referred to as the "knowledge economy." The technologies that assist in providing vital goods and services (including food and health care) of uniform quality to huge masses of people require workers with some very specific information skills — skills that need to be updated on a regular basis. Since there is a shortage of appropriate people and since even those with the right qualifications have to work hard to remain current, a hefty process of education (and/or training) is required to ensure that roles are filled.

What is of concern for the humanistic educator is that the globalizing forces and the perceived need for the creation of a new type of workforce are leading to the deepening of economic gaps and the destruction of languages and cultures (Petras and Veltmeyer, 2001; Hoppers, 2000; Hoogvelt, 1997; etc.), along with changes in funding structures that have an impact on social justice and equity (e.g., Medina, 2001; Porter and Vicovich, 2000). Acknowledging that this argument is valid, it can still be proposed that viewing technology-based educational projects from an alternative perspective will provide an opportunity to consider steps to lead to consequences that are less destructive. It is also worth considering whether the potential benefits of these new opportunities can, in fact, strengthen diversity. Equitable broadband access and the creation of funding structures to accomplish this would help to overcome the gaps that are being made wider by globalizing forces. For example, in the Canadian Broadband Internet Survey (Tewanee, March 2002, personal communication), it was shown that in order to manage their lands, First Nations need access to GIS and other computer-based applications, many of which are available via the Internet, yet require high speed to download or transmit. However, remote communities have no infrastructure to make this a reality.

The Futures Project (2000) looks at how distance education has had a role in the developing world and describes how information technologies can alleviate the "brain drain" and open doors for isolated groups of people. Edwards (2002) suggests that alongside of certain types of homogenization, the trend of global-ization also "produces a pressure for local autonomy and identity which may be asserted in a range of ways" (p. 103). Indigenous distributed learning projects confirm this, with community needs assessments and the preservation of heritage and language often appearing as key components in the planning process. Ideally, through appropriately designed learning environments, tradi-tional cultural values can be synthesized with skills related to modern job requirements (e.g., Spronk, 1995).

From a more mainstream point of view, in the context of currently changing paradigms supporting university effectiveness (the subject of this book and a term that can and will be explored further), the predominant indicator of globalization appears to be the provision of standardized educational experiences to ever increasing numbers of learners. Coalitions have begun to form (such as Universitas 21[3]) and are examples of postsecondary course delivery partnership strategies that cross national boundaries. Another related trend is an increasing focus on science and technology as the most valued subject areas, with a decreasing emphasis on liberal arts (e.g., Blackmore, 2000; Walters, 2000; Carnoy, 2002; etc.). Going global in education generally implies the use of distributed learning technologies for geographically widespread groups of learners and, in the view of efficiency, this idea often translates into a homogenization of educational content. To create distributed learning programs that are marketable to as wide of an audience as possible tends to become the objective because it is seen to be profitable. This is particularly true when education becomes either a privatized commodity or delivered by a public institution suffering from severe, ongoing budget cuts. Both funding scenarios are no doubt familiar to readers of this volume and both diminish the view of education as a necessary process to be supported for the public good. With the profit motive at the forefront, multinational corporations have for some time been providing everything from entertainment to running shoes to soft drinks in a cheaply produced form for mass consumption. In the same way, technology coupled with the privatized (or pseudo-privatized) concept of educational institutions can now easily allow an educational institution to provide generic learning experiences to vast audiences (or markets) who are not geographically close and who may have widely different cultural backgrounds. If the economic imperative becomes dominant in education, what is the social impact? To what extent is the social responsibility that has previously defined the purpose of public education now at risk? Does the combination of technology and the free enterprise global market have to imply a "one-size-fits-all," homogenized approach to learning? Or, can a vision evolve where appropriate educational content will be developed that is customizable to meet diverse needs? If so, who will determine the standards and how can quality be measured?

The Issue of Quality in "Technology-Enhanced" Global Education

Quality in industry generally refers to the maintenance of consistent standards. In education, the concept of quality often translates into learning being measured by tools such as standardized testing. Debate on this topic rages, with a main concern being whether available tools do a good job of measuring what is of value in education, or whether the requirement for measuring actually ends up driving what is taught. With globalization and distributed learning creating a need for educational credentials to cross boundaries in ways that have not been necessary in the past, quality is a hot topic.

In a case study of Australian higher education focusing on how globalization has had impact on gender equity in the academic domain, Blackmore (2000) begins by stating: "Globalization has been the justification for the radical transformation of education in the late 1990s in most Western capitalist states. We are told that we must be more flexible, work harder, and develop our technological skills in order to make education contribute more to national productivity and to achieve international competitiveness" (p. 333). She continues with an analysis of how privatization has been seen as the solution to the reduced government funding that is part of the global competitive emphasis. This, in turn, has led to a quality movement that focuses on teaching and minimizes research, has allowed a new ethos of corporate influence on educational content, and has commodified the curriculum with "well-packaged," but generic (and perhaps even teacherless) online courses that are often based on behaviorist instructional design principles.

The issue of academic quality is discussed in other chapters of this volume. In this chapter, issues relevant to how or whether diversity might coexist with the concept of accountability and standards will be addressed. The question of what makes an effective university in an internationalized higher education context was dealt with in a recent presentation for the Council for Higher Education Accreditation (CHEA) international seminar (Van Damme, 2002). Van Damme noted that internationalization is not the same as globalization and stated that "globalization now seems to be a much more appropriate concept to come to terms with changes in the higher education sector."

Van Damme also refers to protective and protectionist national policies and calls for a global regulatory framework to ensure quality and discourage rogue providers. To do otherwise is to risk "globalization and marketization without restraints" (p. 6). He states that "real competition only is possible if consumers can found their decisions on explicit and reliable information" (p. 10). Who would determine international standards? Van Damme describes a range of strategies to develop meta-accreditation and suggests several strategies, including reference to the SEEL Quality (e)Learning in a Knowledge Economy and Society conference in Lisbon (May, 2003). This society (Supporting Excellence in E-Learning) is a consortium dedicated to the study of the impact of quality policies in e-learning at local and regional levels. "Regions are becoming focal points for knowledge creation and learning in the new age of global, knowledge-intensive capitalism, as they in effect become learning regions. The learning regions function as collectors and repositories of knowledge and ideas, and provide the underlying environment or infrastructure which facilitates the flow of knowledge, ideas and learning" (SEEL, 2003). This view may well provide a balance between local community-based needs and globalized sharing. Clearly there are the beginnings of important work being done in this area.

Medina (2001) describes how in the face of the "growing tendency to borrow market practices in order to implement them in university policies" (p. 37), wealth is becoming more important than liberal education, corporate influence is growing and is having an impact on educational policy, intellectual property is more closely protected by academic institutions[4] and in education as elsewhere, globalization means that autonomy is potentially being taken away. In this case it is faculty who, if they wish to retain autonomy, must work through "research, teaching and political commitment ... [for] ... both justice and empowerment" (p. 37). As a balancing force, Medina references the Council on International Education Exchange (CIEE) whose mission is "to help people gain understanding, acquire knowledge and develop skills for living in a globally interdependent and culturally diverse world[5]."

The view of university effectiveness must incorporate a vision of the type of society that is ultimately desired and one of the most complex considerations in any educational endeavour is the role of culture and diversity. The use of globalizing technologies increase this complexity and when a financial bottom-line becomes a priority, it is easy to neglect issues of social impact. Blackmore (2000) describes this as encouraging "'immediate use value' rather than thinking about curriculum development based on professional and academic judgment" and describes how critical thinking and inclusiveness are declining, along with the demise of pure foundational disciplines such as anthropology and history (p. 345) as a utilitarian view of education gains in strength. Evaluation methods related to the quality movement have an impact here as well, as "teaching gets substantively harder as the capacities to attract, retain and pass students become key performance indicators, course requirements (e.g., advanced standing, length of courses, etc.) have been lowered, and international students have required greater language skill support" (p. 346).

Glocalization

In contrast to terminology reflecting a centralized and homogenous view of efficiency and quality, Edwards (2002) and others (e.g., Marshall and Gregor, 2002; The World Bank Institute, etc.) use the term "glocalization" to describe an educational resource development process that includes a local development team partnered with a centralized institution. Glocalization incorporates learning materials from both the local community and the vast collection of worldwide online sources. In this view, advanced communications technologies can reconfigure, rather than supplant, cultural diversity. The goal is to create a "glocal" learning environment, incorporating local knowledge and culture, but also linking learners internationally and, as one example of how this is being addressed, the World Bank Institute has initiated a "Glocal Youth Parliament"[6] which held its first annual conference in 2002. There is scope for a much more in-depth body of literature to develop on this topic. The focus is on how communities can work in ways to benefit from the advantages that a globalized tool such as distributed learning can offer without succumbing to generic or non-culturally relevant learning experiences. In doing so, communities will be strengthening their abilities to interact in the broad worldwide arena if and when they so desire, while retaining unique identities and cultural heritage.

Indigenous Pilot Projects and the Diversity Discourse

In 2002, Royal Roads University (RRU) in Victoria, British Columbia, Canada, launched two new pilot programs that focused on Indigenous learners. Much admiration and thanks is given to the learners, elders, instructors, First Nation Communities in BC, Corporate Sector sponsors and contributors that included representatives from RRU and various levels of governments that participated and contributed to the programs. Congratulations are also extended to all of the graduates of the programs that are now applying their applied assignments. Both programs were delivered through distributed learning and included on-campus residencies, as well as online learning experiences. Based on consultation and research, the topics for these programs were Indigenous Corporate Relations (ICR) and Distributed Learning Facilitation for Indigenous Communities (DLFIC).

The focus was to build upon already existing capacity building initiatives with First Nations and aboriginal communities in, for example, education, language preservation, economic development and health. Both programs were designed to, and ultimately did, include learners and instructors whose home communities were geographically dispersed — incorporating a range of remote and urban locations within western Canada, both on and off reserve.

The literature review section of the above report described the following key points:

- There are many projects indicating that distance education technologies can be effective in Indigenous communities.

- Appropriately designed learning environments that include onsite facilitation can combine traditional cultural values with skills related to modern job requirements and to the furthering of an academically oriented education.

- Distributed learning environments can provide flexibility and self-pacing to meet the varying learning needs of learners within Indigenous communities and can also avoid the negative aspects that centralized institutional learning environments often have for these learners.

- Community consultation and needs analyses within communities are important factors to ensure that distributed learning projects are effective.

Results showed that distributed learning was seen as a good fit to meet community needs because it provided the flexibility of allowing learners to experience educational opportunities without leaving community responsibilities. The history of colonization also was a factor to be considered in terms of residual negative views of educational institutions and reluctance to participate in mainstream education (e.g., see the Royal Commission on Aboriginal Peoples, 1996). Aspects of the classroom-based, institutional/industrial educational model have been acknowledged by many writers and researchers to have been particularly damaging to Indigenous communities and there are background and educational issues that need to be understood to effectively design new and relevant programs (e.g., see Spronk, 1995; Distributed Learning Task Force, 1997; Davis, 2000). Based on one of the author's experiences, successful Aboriginal programs, particularly delivered at a community college and university level, that meet the need of building local community capacity exist and are successfully delivered in a classroom. However, these programs that want to expand and utilize educational technologies are located within non-aboriginal institutions and face constant upward challenges, specifically over financial, land issues and institutional priorities towards potential global expansion. The authors of this paper hold the view that institutions that do not fully acknowledge aboriginal title and rights at a community level prior to developing global expansion plans are creating a wider digital divide.

The RRU pilot programs tried to emphasize the need for larger institutions to understand the very underpinnings of why a digital divide exists between remote and Indigenous communities (the "have nots") and mainstream societies and educational institutions (the "haves"). RRU has a lot to offer with a wide range of programs delivered via distributed learning, but in conversation with many aboriginal people, you will hear the institution referred to as the "epitome of imperialism." With its defining architectural heritage being a castle that was built to be home to an elite British Colonial family during a time when the local

Indigenous people suffered appalling consequences from colonization, its function as a military college for 40 years and its current involvement in contracts with government in treaty and land use studies and assessments, it is perhaps not surprising that RRU would be viewed in this way. Breaking down the perceived walls (silos) of imperialism at RRU required the understanding of deeply entrenched behaviour that needed to change in order to provide a foundation to assist both groups in building capacity to work together in a globalized world. Much work is still required.

These underpinnings include the very different world views, assumptions and beliefs that define and provide identity for remote and Indigenous communities and conflict with mainstream western society's world views, assumptions, values and beliefs. More often than not, these very underpinnings further polarize mainstream communities and remote and Indigenous communities and cause culture clashes between the two. Such culture clashes have caused discord in relationships involving education and Indigenous communities. Also in this mix are the historical relationships that mainstream education has been involved in, including offering little or no recognition for traditional means of education via traditional knowledge keepers, heads of families and other pre-contact governance systems both orally and in the written form that occurred in Indigenous communities prior to European contact. These underpinnings can also be categorized in the evolution of delivery of education as follows (Claeys & Wellburn, 2000):

- *Pre-contact* — education delivered by knowledge keepers and based on traditional philosophies of Indigenous communities.

- *Contact* — education delivered by missionaries.

- *Post-contact* — education delivered via residential schools and via colonial, government-controlled systems.

- Education delivered by assimilation into mainstream systems, which were not based on constructivist models of collaborative education.

- And, finally, today where Indigenous communities are empowering community members to take control of local education via face-to-face and distributed models that are inclusive of the different world views, assumptions and beliefs of both western culture and Indigenous people.

Without the understanding of these underpinnings, both mainstream society and Indigenous communities will never fully engage and allow for harmony to occur in partnership building in a local educational or in the global educational context.

Partnerships in the Pilot Programs

Partnership building for the pilot programs also included planning via outreach consultation (meetings across Canada) and an advisory committee where issues for planning and developing the programs were brought to the table for consideration. The consultation process was also included into the foundations of the two programs as applied assignments for learners from participating communities. This core value is seen as essential in building partnerships for education. Many first nation communities also signed Memoranda of Understanding with RRU to further develop partnerships for this purpose.

Ultimately, the programs were implemented following the Royal Roads model with its cohort-based groupings for ongoing collaboration, its combination of short residencies and longer periods of distributed learning experiences, and its outcomes-based and applied assignments.

Partnerships with communities were key to both programs and the emphasis was on solving real community problems. In ICR, learners worked to create new strategies for relations between corporations and Indigenous communities. Corporations have views similar to what Walters (2000) refers to as a form of "risk management"[7] that has become necessary due to global competitiveness. ICR learners looked at creating mutually beneficial agreements between corporations and local communities that could best accommodate the needs of both and move towards greater implementation of aboriginal title and rights, specifically in the natural resource sectors. The ICR program was also based on recent Canadian Supreme Court rulings that support aboriginal title and rights within an Indigenous Corporate context that opposed corporate sovereignty over natural resources. In DLFIC, projects were designed where the final outcome was the implementation of a learning center or community learning program that met needs identified through the needs assessment process that had been undertaken as an early program assignment.

Note that in both programs the resulting topics for major projects were extremely varied from learner to learner based on individual relevancy to their situations. The range of project types added a layer of difficulty to measuring what was learned, but since a variety of projects had been anticipated, the learning outcomes had been developed to a large extent with a focus on having the learners demonstrate that they were able to implement processes (such as analyzing community needs) and think critically about issues. With care, these types of outcomes were possible to demonstrate and measure, even in projects that appeared to be widely disparate.

Parallel to some of the other difficulties identified in the consideration of globalization of education with respect to quality and standards, some interesting issues emerged with respect to the Royal Roads Indigenous programs. Academic culture is based on premises that do not always mesh easily with traditional learning in Indigenous communities and points of contention included such concerns as entry requirements, credentialing, qualifications required for instructors and the time frames in which course assignments could be completed by learners who were involved in community activities. These issues were not always easily resolved and it became clear that resolution requires the necessary commitment and resources to fully address emergent issues.

Another key issue related to the use of Indigenous knowledge (IK) within the programs. It was made clear at the onset that projects created by learners in the program would not require the learner or the learner's community to have to give up ownership of any traditional knowledge that was brought into the process. This empowering strategy is in alignment with thinking that was shared at a recent conference titled, "World's Indigenous Peoples: Perspectives and Processes" (which took place in Kelowna, BC, Canada, October 2002), referencing current literature such as Bell (2000) and Brascoupé and Mann (2001, p. 1) who state, "Too often, IK issues are examined from the perspective of researchers and policy makers. Glaringly absent is a community perspective that focuses on community control and management, even though it is the communities, no one else, that is responsible for guarding and transmitting this knowledge."

That community knowledge was to remain community-based and that each community was expected to participate in the role of overseeing how such knowledge would be used set a precedent in how courses have been delivered at Royal Roads University. This concept of a partnership may be an extremely important consideration for other institutions that want to "go global," but also want to create learning experiences that are not homogenized and generic. A topic for discussion is the issue of intellectual property vs. Indigenous knowledge and the already entrenched, global policies regarding the difference.

Dighe (2000) states, "The discourse on diversity not only sees participant groups as capable of development on their own, but also sees them as being already engaged in daily processes of self and community development. Rather than outside 'experts' planning and designing programmes for meeting the needs of different groups, the diversity discourse allows for great decision-making on the part of the participant groups. Educational programmes emerging from this discourse would thus look for increasingly diverse solutions rather than propagating universal solutions for all."

For Indigenous distributed learning projects, questions such as: *"How can appropriate partnerships be established? How can the technologies that support global communication be used to meet community needs and build capacity? Who will decide? How should learners receive credentials?"*, have a particular relevance that needs to be further explored.

Strategic partnerships sought in the design and delivery of education by graduates of the Royal Roads DLFIC program would have graduates in their respective communities forming alliances between mainstream educational institutions and Indigenous communities that would allow for an equal transfer of knowledge between the two. These alliances or affiliation agreements may include a collaborative model of delivery of education, combining the strengths and weaknesses of both systems in delivering onsite education to Indigenous communities. These alliances and affiliations sought in the Royal Roads pilots could potentially address communication disorders and even perhaps develop assistive technologies and software, specifically in Indigenous languages, that may address barriers in education for Indigenous peoples and vise versa.

The two Royal Roads pilot programs with their limited resources also tried to provide due diligence in these following areas:

- Address the geographical isolation of First Nation Metis and Inuit people in the creation of learning opportunities which are facilitated by distance education delivery modes;

- Address the under-representation of Indigenous peoples in educational institutions;

- Develop courses, curriculum and strategic planning related to Indigenous topics and education;

- Acknowledge Indigenous culture, identity and pedagogy; and

- Incorporate ethical research practices.

Notwithstanding the fine intentions of the programs, it has been learned that essentially it is more difficult to work within contrasting world views without the trust and commitment of all parties involved than it would be to work exclusively in one or the other.

Humanizing Globalization (A Conclusion)

It is perhaps ironic that as instant communication and information transfer allows us to know more about each other than has ever been the case before, many people and cultures feel more threatened than ever, conflict and crises such as war and terrorism are far from having been eliminated and we don't really feel connected in a global way. It is possible that a competitive nature in education is a factor contributing to the problems of how we get along. Many traditional Indigenous educational views support the concept that a less individualistic and more community-focused form of education is a more sustainable model. The authors of this paper look for ways that distributed learning can be developed to enhance collaborative learning, possibly even across diverse cultures, in ways that will allow for respectful connections (e.g., Bates, 1995; Farrell, 1998; Wellburn, 1999; Claeys and Wellburn, 2000). Certainly within the Indigenous distributed learning projects there was a sense of the possibility of strengthening capacity through maintaining a community, while tapping into a vast global network. It is not an easy task, but to critically evaluate the situation and work with the positive potential seems to be a better response than accepting globalization as a homogenizing force that cannot be stopped or raging against it in hopes that it will go away.

References

Barro, R. and Lee, J. (2000). *International data on educational attainment: Updates and implications.* Retrieved March 18, 2003 from the World Wide Web: http://www2.cid.harvard.edu/cidwp/042.pdf.

Bates, A. W. (1995). Creating the future: Developing vision in open and distance learning. In F. Lockwood (Ed.), *Open and Distance Learning Today.* London: Routledge.

Battiste, M. (2000). *Protecting Indigenous Knowledge and Heritage.* Vancouver, BC, Canada: UBC Press.

Bell, C. (2000). Protecting indigenous heritage resources in Canada: Kitkatla vs. B.C. *International Journal of Cultural Property,* 10, 246-263.

Blackmore, J. (2000). "Hanging onto the edge": An Australian case study of women, universities and globalization. In N. Stromquist and K. Monkman (Eds.), *Globalization and Education: Integration and Contestation Across Cultures.* MD: Rowan & Littlefield Publishers.

Brascoupé, S. and Mann, H. (2001). *A community guide to protecting Indigenous knowledge. Research and analysis directorate.* Retrieved March 13, 2003 from the World Wide Web: http://www.ainc-inac.gc.ca/pr/ra/ind/gui_e.pdf.

Brown, J. S., Collins, A. and Duguid, P. (1989). Situated cognition and the culture of learning. *Educational Researcher,* 18, 32-42.

Carnoy, M. (2000). Globalization and educational reform. In N. Stromquist and K. Monkman (Eds.), *Globalization and Education: Integration and Contestation Across Cultures.* MD: Rowan & Littlefield Publishers.

CED (2002). *A shared future: Reducing global poverty*. Retrieved March 18, 2003 from the World Wide Web: http://www.ced.org/docs/summary/summary_globalization.pdf.

Claeys, G. and Wellburn, E. (2000). *Partnerships and perspectives in distributed learning for indigenous communities*. Retrieved March 23, 2003 from the World Wide Web: http://www.royalroads.ca/resources/progressrpt.pdf.

Davis, L. (2000). Electronic highways, electronic classrooms. In M. Castellano, L. Davis and L. Lahache (Eds.), *Aboriginal Education: Fulfilling the Promise*. Vancouver, Canada: UBC Press.

Dighe, A. (2000). *Diversity in education in an era of globalization*. Retrieved March 12, 2003 from the World Wide Web: http://www.learnder.org/dl/VS3-00q-Diversity.pdf.

Distributed Learning Task Force (BC) (1997). *Access and Choice: The Future of Distributed Learning in British Columbia*. Victoria, BC, Canada: Centre for Curriculum, Transfer and Technology.

Edwards, R. (2002). Distribution and interconnectedness: The globalisation of education. In M. Lea and K. Nicoll (Eds.), *Distributed Learning: Social and Cultural Approaches to Practice*. New York: Routledge Falmer.

Egea-Kuehne (2003). *The commodification of education: Ethico-political issues in the global marketing of knowledge*. (Paper to be presented at The Learning Conference: Institute of Education, University of London, July 2003). Retrieved March 22, 2003 from the World Wide Web: http://learningconference.com/ProposalSystem/Presentations/P000295.

Farrell, G. (1999). *The development of virtual education: A global perspective*. Retrieved September 12, 2002 from the World Wide Web: http://www.col.org/virtualed/.

The Futures Project (2000). *The Universal Impact of Competition and globalization in Higher Education*. Retrieved March 12, 2003 from the World Wide Web: http://www.futuresproject.org.

Hoogvelt, A. (1997). *Globalization and the Postcolonial World: The New Political Economy of Development*. Baltimore, MD: John Hopkins University Press.

Hoppers, C. (2000). Globalization and the social construction of reality: Affirming or unmasking the "inevitable"? In N. Stromquist and K. Monkman (Eds.), *Globalization and Education: Integration and Contestation Across Cultures*, MD: Rowan & Littlefield Publishers.

Lowi, T. (2000). Think globally, lose locally. In G. Lachapelle and J. Trent (Eds.), *Globalization, Governance and Identity: The Emergence of New Partnerships*. Montreal, Canada: University of Montreal Press.

Marshall, S. and Gregor, S. (2002). Distance education in the online world: Implications for higher education. In R. Discenza, C. Howard and K. Schenk (Eds.), *The Design & Management of Effective Distance Learning Programs*. Hershey, PA: Idea Group Publishing.

Medina, A. (2001). *The impact of globalization on higher education.* Retrieved March 12, 2003 from the World Wide Web: http://www.umbc.edu/11c/PDFfiles/theimpactofglobalization/.pdf.

Mussa, M. (2000). *Factors driving global economic integration.* Retrieved March 13, 2003 from the World Wide Web: http://www.imf.org/external/np/speeches/2000/082500.htm.

Petras, J. and Veltmeyer, H. (2001). *Globalization Unmasked: Imperialism in the 21ˢᵗ Century.* London: Zed Books.

Porter, P. and Vidovich, L. (2000, Fall). Globalization and Higher Education Policy. *Educational Theory*, 50(4), 449-466.

Radcliffe, S., Laurie, N. and Andolina, R. (2002). *Indigenous people and political transnationalism: Globalization from below meets globalization from above?* Retrieved March 13, 2003 from the World Wide Web: http://www.transcomm.ox.ac.uk/working%20papers/WPTC-02-05%20Radcliffe.pdf.

Royal Commission on Aboriginal Peoples (1996). *Report of the Royal Commission on Aboriginal Peoples*. Ottawa: Ministry of Supply and Services.

SEEL: Supporting Excellence in E-Learning (2003). Quality (e)Learning in a Knowledge Economy and Society. Retrieved March 23, 2003 from the World Wide Web: http://www.seelnet.org/seel/.

Spronk, B. (1995). Appropriate learning technologies: Aboriginal learners, needs and practices. In E. M. Keough and J. M. Roberts (Eds.), *Why the Information Highway?: Lessons from Open and Distance Learning*. Toronto: Trifolium Books.

Stromquist, N. (2002). *Education in a Globalized World: The Connectivity of Economic Power, Technology and Knowledge*. Oxford: Rowman & Littlefield.

Stromquist N. and Monkman, K. (2000). Defining globalization and assessing its implications on knowledge and education. In N. Stromquist and K. Monkman (Eds.), *Globalization and Education: Integration and Contestation Across Cultures*. MD: Rowan & Littlefield Publishers.

Tabb, W. (2001) *Essay: Globalization and Education as a Commodity.* Retrieved March 18, 2003 from the World Wide Web: http://www.psc-cuny.org/jcglobalization.htm.

Van Damme, D. (2002). *Quality assurance in an international environment: National and international interests and tensions.* Retrieved March 13, 2003 from the World Wide Web: http://www.utwente.nl/cheps/documenten/Susdamme.pdf.

Walters, S. (2000). Globalization, adult education and development. In N. Stromquist and K. Monkman (Eds.), *Globalization and Education: Integration and Contestation Across Cultures.* MD: Rowan & Littlefield Publishers.

Wellburn, E. (1999). Educational vision, theory, and technology for virtual learning in K-12: Perils, possibilities, and pedagogical decisions. In C. Feyten and J. Nutta (Eds.), *Virtual Instruction: Issues and Insights from an International Perspective.* Englewood, CO: Libraries Unlimited.

The World Bank (2000). *Poverty in an Age of Globalization.* Retrieved March 13, 2003 from the World Wide Web: http://www1.worldbank.org/economicpolicy/globalization/documents/povertyglobalization.pdf.

Endnotes

[1]   See Radcliffe, Laurie and Andolina (2002) for a very different perspective on how globalization impacts Indigenous cultures.

[2]   See Brown, Collins and Duguid (1989) for an outline of the concepts of cognitive constructivism, in particular the concept that knowledge does not exist separately from the culture in which it is situated.

[3]   Universitas21 is described at http://www.universitas21.com/.

[4]   (and thus less likely for ownership of knowledge to be retained by a community and more likely for it to become "commodified") See the section in this chapter on Indigenous Knowledge. Note that some widely-held mainstream concepts of intellectual property are seen by many Indigenous communities as being counter to a view of the sacredness of certain types of knowledge and that since intellectual property agreements often end up with knowledge ultimately residing in the public domain this is especially problematic and in fact can be devastating to cultural re-connectivity.

[5]   See http://www.cie.org.

[6]   Details of the Glocal Youth Parliament are available at http://www.glocalyouth.org/.

[7]   "Flexibility across boundaries of established knowledge domains is increasingly being accepted globally as necessary as new economic and social

problems are addressed. Global competitiveness is driving a growing number of corporations around the world to adopt approaches to 'risk management' that require them to have more holistic and integrated approaches to adult education and training. For example, in remote mining areas in Canada and Chile, companies are 'managing risks' and these include working with Indigenous people and managing the environmental impact, in order that there be social and environmental sustainability. This leads to agreements with local populations, which build on Indigenous knowledge and incorporate some of the best participatory practices of community adult education to engage in an integrated approach to community development. These corporations recognize the need to work closely with the local populations in order most effectively to succeed in maximizing profits" (Walters, 2000, p. 206).

Section II

Course Development
Instruction and
Quality Issues

Chapter V

Online Course Design Principles

Lance J. Richards
Texas A&M University, USA

Kim E. Dooley
Texas A&M University, USA

James R. Lindner
Texas A&M University, USA

Abstract

*The premise of this chapter is that technology for course delivery will change, but effective delivery of content is dependent upon use of appropriate instructional design techniques. The authors take a practical approach by providing guidelines for designing online courses and programs. These guidelines include: (1) designing or selecting a course management tool, (2) course planning and organization, (3) "chunking" content, (4) using interactive teaching and learning strategies, (5) applying adult learning principles, (6) considering self-directed and student-centered learning approaches, (7) using authentic assessment strategies, (8) providing online orientation and technology training, and (9) providing information about appropriate infrastructure for learner support. We use a graduate course, **Advanced Methods in Distance Education**, as "the case" to provide specific examples of the instructional design components. By following these approaches, you can develop a successful online learning environment.*

Introduction

Areas of competence important for teaching at a distance include course planning and organization, verbal and nonverbal presentation skills, collaborative teamwork, questioning strategies, subject matter expertise, involving students and coordinating their activities at field sites, knowledge of basic learning theory, knowledge of the distance learning field, design of study guides, graphic design and visual thinking (Cyrs, 1997). Purdy and Wright (1992) asserted that, "it is not that the technology underpinning distance education drives the system, but rather that fundamental changes in teaching style, technique, and motivation must take place to make the new 'classrooms' of the present and future function effectively." What fundamental changes must instructors make to make distance learning more effective and appropriate for a growing audience?

Often organizations focus on the technological infrastructure to build distance education programs without giving regard to the importance of instructional design. Technology will change — satellite, interactive video, Internet, CD-ROM — but effective delivery of content will remain dependent upon appropriate instructional design techniques.

Newcomers to online instruction find that instructional design principles are very different for this medium. Principles that worked in a face-to-face environment or even over video/videoconferencing must be modified to facilitate online learning. Issues of social presence and immediacy behaviors are extremely important (Gunawardena and Zittle, 1997) and the role of the instructor as a facilitator/coach is critical. Now more than in traditional classrooms, distance education relies upon the student's ability to be self-directed and motivated (Lindner and Murphy, 2001).

This chapter provides practical guidelines to designing online courses. By following these steps, you can develop a successful, active, online learning environment. We will use *Advanced Methods in Distance Education*, a graduate course, to illustrate each of these design principles.

Origins of a Course

In the past, *Advanced Methods* has been taught using a combination of interactive video and Web Course Tools (WebCT®), but we recently redesigned the course completely for asynchronous delivery. We wanted to make the course learner-centered and competency-based (Lindner and Dooley, 2002), rather than relying on more traditional "contact hours." The course includes five modules based upon the core competencies needed by the distance education professional (Williams, 2003; Dooley and Lindner, 2002; Thach and Murphy, 1995). We selected a house to illustrate the entire bundle of competencies, with various rooms representing the module themes (see Figure 1). The five course modules and their respective locations in the house include the basement, *Foundations of Teaching and Learning at a Distance*; the workroom, *Technology Knowledge and Skills*; the kitchen, *Instructional Design*; the den, *Adult Learning Theory*; and the study, *Administrative Issues*.

Another component of our course design is measuring competence in the modules. At the beginning of the semester, each learner completes a self-assessment instrument that corresponds with the core-competency areas in the course. A learner who already demonstrates expertise in certain areas can complete the outcome assessment measures in that module and move on to the next module. In modules where the learner lacks expertise, he or she can spend more time building competence. Our competency-based approach allows the students to move through individual lesson sequences and allows the instructor to act as a facilitator and authenticator of competence. At the end of the semester the instructor uses an evaluation rubric to authenticate the students' competence in designing a lesson for distance delivery. The learner also completes the self-assessment instrument again to document growth in each of the core-competency areas. The learner should be self-directed and motivated to excel in this adult learning/asynchronously delivered environment (Lindner, Dooley and Murphy, 2001; Grow, 1991).

Figure 1.   Advanced Methods Main Course Interface Showing Module Design

Within the framework of our case study of *Advanced Methods* as a course template, we can demonstrate important online-course design principles.

Design or Select a Course Management Tool

Design or select your course management tool to maximize the efficiency of course management by the instructor and ease of use by the students. Many universities are using Web course tools, such as WebCT® or Blackboard®, to facilitate the online instruction process because of the many features that are made available. These features include: an electronic grade book, password protection for both course and student data, communication tools, tools for managing student assignments, and license and support by the university. Research has shown that the use of Web course tools contributes to students' abilities to accomplish course objectives (Lindner and Murphy, 2001). Whatever course management tool you choose should be user-friendly and intuitive for the learner, as well as accommodating to a variety of different learning styles.

WebCT® is the course management tool for *Advanced Methods* and serves as the access point for students. Figure 2 shows how WebCT® can link course content designed in modules with communication tools, course assignments, team projects, and student grades.

Figure 2. WebCT® Interface Used in Advanced Methods

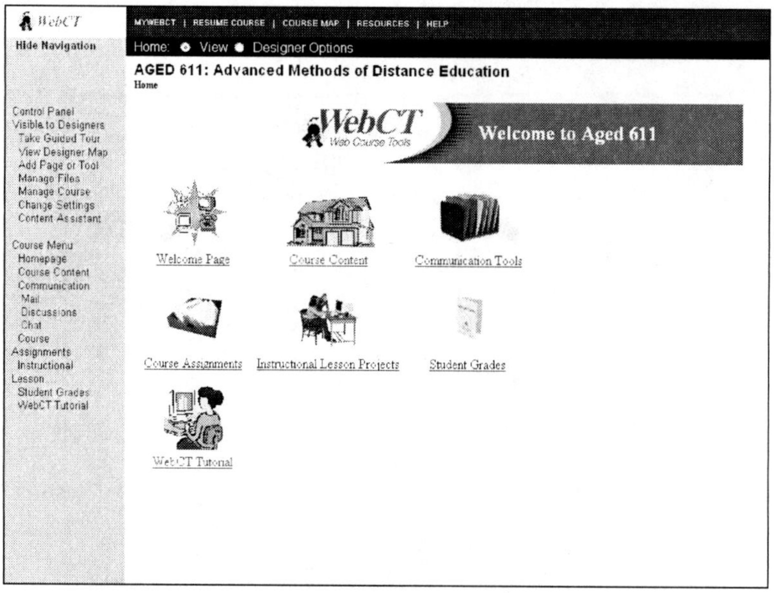

Course Planning and Organization

It is never enough to simply save text as HTML and assume you have developed an online course. People tire of reading on a computer or television screen and will end up printing it out. If your only objective is to deliver static text, you might as well send a printed manual rather than using expensive telecommunication technologies. (For a comparison of print versus online media, see Figure 3.) A well-planned electronic course design will begin with storyboards to create a visual map indicating the placement of intuitive icons, readable text formatted for computer or TV display, good use of color and clip art or pictures, and perhaps animated graphics to enhance understanding.

When delivering online, strive to give courses a common look and feel. Common course components include:

Course syllabus. This should include, but not be limited to, a brief course description, time requirements, information about course delivery, overall course objectives, required textbooks, required meeting times for the course, the grading policy, and a learning outline, including all due dates for the semester or period of instruction.

Course orientation. This section should include a brief biography of the instructor and support staff associated with the course or program of instruction, student expectations for the course, specific information and training concerning the submission of assignments and examinations, links to order the course textbook(s), links to download the plug-ins necessary to access all course materials, technology training, and information about and links to campus support resources for distance education.

Course content. This section should include links to modules and lessons included within the course. Within each lesson, the student should find learning objectives, a list of required readings and activities, supplemental readings and activities for enrichment, and links to multimedia, such as videos or audio-annotated lectures (such as a PowerPoint® Producer file format).

Course calendar. This section should include dates for the beginning and ending of each module, due dates for all assignments, projects, and examinations and university holidays affecting student progress through the course.

Site map. This section provides an alternative way for students to access course information. Instead of navigating through links that are three deep, the site map provides students a "quick-click" means to access all course material from one location.

Index. This section provides an index of and links to all reading material, videos, etc., arranged in alphabetical order (similar to the index in a textbook, but hyperlinked).

Chunk Content

Electronic content should not be delivered in hour-long lectures. Whether teaching online or through interactive television (satellite or compressed video), learners must be given smaller chunks of content. Ideally, plan for about 10 to

Figure 3.   Printed Media vs. Online Media

Printed Material vs. Online

The debate over whether to put all course material online or provide print copies to students is highlighted by the following advantages and disadvantages of print and online.

- *Convenience.* In general, print material is a more convenient medium than online. You can take a book anywhere ... but lugging a computer to the beach is difficult at best.
- *Emotional experience.* Have you ever heard about anyone coming back from vacation saying, "I took my laptop and read some really good articles online while I was away?"
- *Timeliness.* Printed text is frozen at the time of publication and materials cannot be easily updated. Online materials can and should be updated continuously.
- *Waiting.* Online materials are available on demand. Have you ever had to wait for a book order to come in? Or for the library to send you a printed copy of a periodical?
- *Costs.* Print materials are designed for purchase. Online materials can usually be viewed for free on a computer. People can reduce printing costs by only printing what they want to print.
- *Links.* While print material suggests online materials to review ... online material provide hyperlinks to take you right to the material.
- *Hazards.* Computer screens may give off harmful radiation and cause carpal tunnel syndrome. Print may cause your hands to get a little dirty. Have you ever looked at your hands after reading the Sunday paper?
- *Interactivity.* The last interactivity I saw from a book was when a flyer for another book purchase fell out ... enough said.
- *Waste.* Online materials take up less space on your bookshelf and drawers than books and other print material ... unless you decorate your office and house with computer equipment.
- *Adaptability.* I have never met a student that enjoyed buying six textbooks only to read two or three chapters from each book. Online materials allow you to "cherry-pick" materials to be read.
- *Organization.* Students prefer to have all reading materials required for a course to be organized in one central local and in general prefer print material over online material that they have to search for.

This material is based in-part on an online article published on the Mouthshut.com web site: http://www.mouthsut.com/readreview/24768-1.html.

Figure 4. Advanced Methods Interface Showing Lessons Within a Module

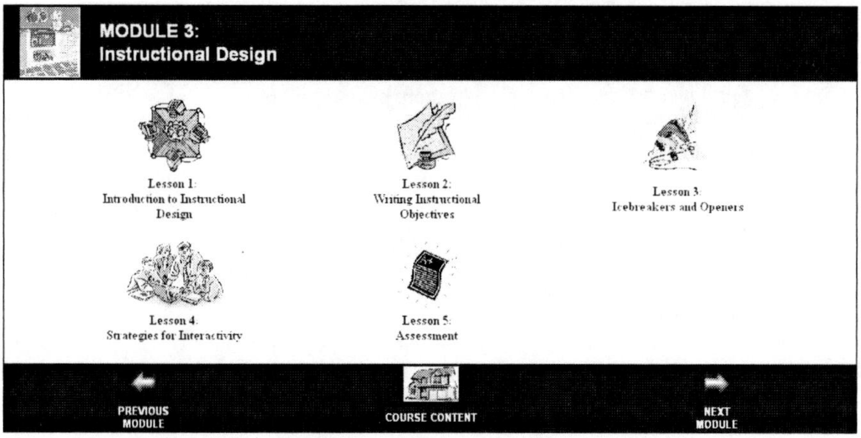

15 minutes and then vary the strategy, shift gears, and change to an exercise or discussion. A variety of methods can be used to facilitate this process, including Adobe Acrobat® printouts, audio-annotated lectures, World Wide Web searches, streaming video clips, interactive Web pages, activities/exercises, chat, whiteboards, and e-mail video discussions. For a graphical representation of chunking content, see Figure 4.

Use Interactive Teaching and Learning Strategies

Many people fear that distance education is not as interactive as face-to-face teaching or training. The amount of interaction we provide in a distance learning environment also contributes to the degree of isolation a student may feel (Shearer, 2003). Online courses can, in fact, be more interactive than face-to-face courses if the instructional design features build in opportunities for interaction with the instructor, other learners, the content, and the delivery tools (media). Web course tools (e.g., static and dynamic Web pages, threaded discussion groups, e-mail, chat, instant messaging, streaming media/video, animations, application sharing, IP audio/video conferencing) are being adopted and used increasingly by teachers to optimize delivery of instructional material (Lindner and Murphy, 2001). Online instruction can also include student exercises with drag-and-drop features, virtual teams, and even a scavenger hunt of Internet resources to build interaction, just to name a few (Figure 5). All it takes is a bit of creativity!

Use technology and the Web's flexibility to create an active learning environment. Learning is a social process, so the interaction with the instructor and other

Figure 5. Increasing Student Interaction in an Online Course: Supporting Deeper and More Meaningful Learning

Shown below are a few examples of activities that online instructors can use to increase student interaction and provide greater opportunity for more meaningful learning.

Icebreakers and openers. Activity designed to ease participants into the course or module of instruction. One example, used by Lani Gunawardena, is a costume party where the students choose a costume representing something about themselves, take a picture of themselves, and post it on the Web with a summary of their personality.

Student expeditions. Give students a list of course topics and have them compile a list of web site references. This is a useful way to acclimate students to the subject matter or the Web technologies used in course delivery. As an addition to this activity students could use quality criteria to rate other students' lists.

PCT (Purposive Creative Thinking). Identify a conflict and locate an analogy basic to the conflict and then produce a solution using threaded discussions or chat.

P2P. Peer to peer interaction where cooperative learning teams share coursework using threaded discussions and chat technology.

Streaming experts. A knowledgeable person provides content to the class via a streaming video presentation or live chat. Students will then continue discussion through threaded discussions or chat.

Mental gymnastics. Student brainstorming on a proposed topic using threaded discussions. Students will rate their peer's ideas and will collectively select the best topic.

Adapted in part from Bonk and Dennen (2003) Table 23.7 and Eitington (1984), *The Winning Trainer.*

students is important. Michael Moore (1989) describes three interactions in learning: learner/learner, learner/instructor, and learner/content. These three levels of interaction "are perhaps more central to what we view as interaction in a distance education course" (Shearer, 2003). Michael Hannifin (1989) itemized the functions that interaction purports to support in an educational context: pacing, elaboration, confirmation, navigation, and inquiry. Interaction will not happen automatically; your role as instructor must shift to that of a facilitator. One method that has worked extremely well is assigning learners to virtual, cooperative learning teams. Each member is assigned to be a facilitator for one of the modules. The role of the facilitator is to promote thought-provoking discussion around the content of that particular module. This builds student/student interaction and provides reinforcement of course content.

Because distance learners often feel isolated, one important aspect of any lesson is gaining attention and stimulating motivation. Use of icebreakers/openers and engaging strategies allow students to take ownership of the learning environment.   In *Advanced Methods*, students are asked to design or modify an icebreaker or opener based on a lesson the student will later deliver to their team members as an asynchronous exercise. The icebreaker/opener is posted to their team folder as a threaded discussion. Students also write clear instructions so their team members can participate in the exercise. The icebreaker/opener is used to build rapport with their team.

Short streaming media files are another way to have presence and immediacy with the learners (Figure 6). Course designers and instructors should consider a variety of ways to build rapport and engage the learners to increase course satisfaction, retention, and interaction.

Adult Learning Principles

Distance education courses draw on the ability of adults to be self-directed. Incorporating adult learning principles (andragogy) into the design and delivery of distance courses will result in more effective learning. Buford and Lindner (2002) describe several adult education principles to include in the development of online courses (as we describe). You can use Figure 6 to ensure adult learning principles (Knowles, Holton and Swanson, 1998) are covered in the design of your courses.

Figure 6.   Example Video/PowerPoint® Presentation Used in Advanced Methods

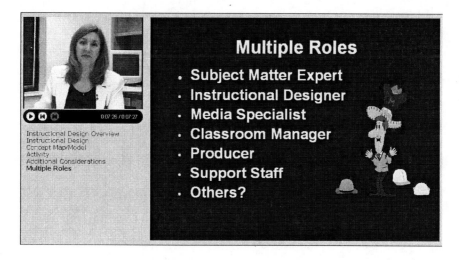

Learner's Need to Know. Adult learners need to know why they need to learn anything before beginning instruction. Your task is to help learners identify gaps between what they know and what they need to know. Self-assessment measures are particularly useful at the beginning of a course to help learners identify gaps in their knowledge, skills, and abilities.

Self-Concept of the Learner. Adult learners want to be in control of their lives. They do not want to be dependent on teachers to "teach" them. Courses should focus on the self-directed nature of learners.

Prior Experience of the Learner. Adult learners come into a course with differing life experiences. They want to use their life experiences to facilitate learning. When possible, focus your teaching on activities that allow learners to draw on and share their prior experiences.

Readiness to Learn. Adult learners learn best when the information to be learned can be directly applied to solving a problem or filling an identified gap. As much as possible, focus on information that can be directly applied to a learner's job.

Orientation to Learning. Adult learners learn better when teaching occurs in the context of real-life situations. Learners are task-oriented in their approach to learning; they want to be able to use acquired competencies to solve real problems. Teaching should be learner-centered, not topic-centered.

Motivation to Learn. Many adult learners want to increase their competencies. They are motivated by internal motivators, such as job satisfaction, self-esteem, and quality of life. To a lesser degree, they are also motivated by external motivators, such as higher pay, better jobs, and advancement opportunities. Develop courses to take advantage of learners' internal motivators.

Comprehensive details on the theory of andragogy are beyond the scope of this chapter. For more information on andragogy and its application, see *The Adult Learner; The Definitive Classic in Adult Education and Human Resource Development* (Knowles, Holton and Swanson, 1998).

Self-Directed and Student-Centered Learning Approaches

Many articles on course design in distance education mention the need for student-centered approaches. Grow (1991) suggests that learners do not necessarily know how to be self-directed. He identifies four levels of self-directed autonomy: dependent, interested, involved, and self-directed. Your goal as teacher, therefore, is to identify the student's level of autonomy and match instructional design to these levels. This process can be used to help you move

Figure 7. Checklist for Adult Learning Principles in Online Course Design

The following questionnaire, adapted from Buford and Lindner (2002), is designed to help teachers develop courses that take advantage of basic principles of adult learning. Read each statement and place a check in the box that most represents your course.

	Rarely	Sometimes	Never
1. Teaching includes structured activities that allow students to identify their competency levels.	☐	☐	☐
2. Students understand the purpose of the teaching activity and why they are participating.	☐	☐	☐
3. The course takes advantage of self-directed nature of learners.	☐	☐	☐
4. The course allows learners to be active participants and not passive observers.	☐	☐	☐
5. The course incorporates real-life examples based on employee experiences.	☐	☐	☐
6. The course content closely matches the learners' interests or content of their job(s).	☐	☐	☐
7. Participation in the course will directly help learners solve problems related to their careers.	☐	☐	☐
8. The course allows students to use their competencies to solve real problems.	☐	☐	☐
9. The course is designed to increase employee job satisfaction.	☐	☐	☐
10. Learners are willing participants in the course.	☐	☐	☐

It is not likely that any one course will be able to take advantage of all adult learning principles. However, the degree to which they can be incorporated into teaching will result in better course design.

learners from dependency toward self-directedness. For example, when you identify the learner as being "interested," you should serve as a motivator and guide, while attempting to move the student to the next level (involved) of autonomy by incrementally shifting your role to that of facilitation. As students become more self-directed, they must be given more autonomy to construct meaningful learning experiences for themselves.

Constructivism is the notion that learners construct their own knowledge from experiences. "Learning from this perspective [constructivist] is viewed as a self-regulatory process of struggling with the conflict between existing personal models of the world and discrepant new insights, constructing new representations and models of reality as a human meaning — making venture with culturally developed tools and symbols, and further negotiating such meaning through cooperative social activity, discourse, and debate" (Fosnot, 1996).

How can you develop courses that meet students' beginning levels of competency and self-directedness and help them grow with other students as the course progresses? Pre-assessment measures are necessary to ensure that you and your students know their competency level (knowledge, skills, and abilities) at the beginning of a course so you can provide instruction targeted specifically for the level of each student. Such information can help students better understand their unique bundles of competencies and increase student satisfaction, motivation, learning and, ultimately, success in a course (Dooley and Lindner, 2002). Competency-based feedback can provide a foundation for student-centered learning plans. This feedback can also be used to describe minimally acceptable knowledge, skills, and abilities on identified core competencies, thus giving instructors tools and information needed to improve curricula, teaching materials, evaluation processes, and instructional delivery methods (Dooley, Lindner and Richards, 2003). A sample from the pre-assessment instrument used in *Advanced Methods* is pictured in Figure 8.

Pre-assessment information can be used to help students progress through a course. Grow (1991) observed that self-directed learning "will rarely be linear, and most classes will contain students at different stages of self-direction." Students should be allowed to progress through the course at different stages. If students demonstrate expertise in one competency area, they should be allowed to complete an outcome assessment for that area and move on to other competency areas that need more development. At the end of the semester, the learner should complete a post-assessment to document growth in each core competency area presented in the course.

Use Authentic Assessment Strategies

One question always asked about distance education courses is, "How do you assess learning?" In the interactive, collaborative environment created by the Internet, we recommend using the three P's: Papers, Projects, and Portfolios. Some courses may require more traditional assessment measures. For this purpose, Web course tools often include online quizzing/testing features. With these tools, the instructor creates test banks and the tools randomly select questions for the learner, grade them, and automatically insert the score in the grade book. Typically these test items have one right answer. They assess knowledge and are indirect indicators of more complex abilities (Huba and Freed, 2000). However, the challenges faced by adults "tend to be those that require the simultaneous coordination and integration of many aspects of knowledge and skill in situations with few right answers." As we shift to learner-centered approaches in teaching, "we should design assessment to evaluate students'

Figure 8. Snapshot of the Self-Assessment Instrument Used in Advanced Methods to Authenticate Student Competency Levels at the Beginning and Ending of the Course

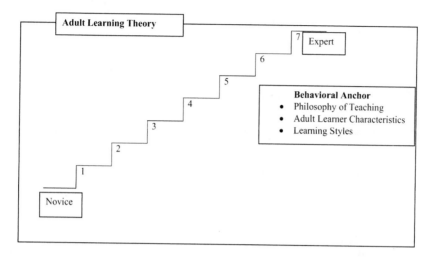

ability to think critically and use their knowledge to address enduring and emerging issues and problems in their disciplines."

We used authentic assessment in *Advanced Methods* to demonstrate mastery of the course competencies by actually walking the students through the instructional design process and assessing their progress along the way. This journey involved the design and development of a 45-minute instructional lesson presented online to members of their cooperative learning teams. The instructor used an evaluation rubric to authenticate competence (Figure 9). Students were checked for mastery in each of the following areas throughout the course:

Lesson topic theme/audience. Students begin the instructional design process by choosing a topic theme for their lesson and defining their audience.

Student-developed instructional design model. Using a constructivist approach, students create a concept map of what they think are critical components of systematic instructional design.

Delivery strategies. Building upon their instructional lesson, students consider the vast variety of instructional media available to bridge communication and facilitate instruction. At this point, students add a list of delivery strategies or support materials they would include in their lesson.

Graphic design. In this section, students create at least two support materials for their lesson. They may choose to create PDF® handouts, PowerPoint® slides with graphics (which may include audio scripting or imported sounds), and links to Internet resources.

Creating an interface for the lesson. At this point, students create a Web page of their instructional lesson, which should simulate an online instructional experience for their team members. Their cooperative learning teams will participate in the lesson, so all necessary components should be available online. Students will include their topic theme and target audience, their lesson objectives and content, their icebreaker and/or opener, some of their interactive strategies, links to support materials, and an outcome assessment.

Lesson evaluation. Students create a lesson evaluation (different from the learner outcome assessment to determine if the instructional objectives were accomplished), using the text and supplemental readings and resources to guide them. Students then participate in each instructional lesson within their cooperative team and complete their team members' lesson evaluations.

Provide Online Orientation and Technology Training

Although most teachers are comfortable with their content and face-to-face presentation techniques, teaching online requires instructors to also be comfortable with the technology tools so that the technology becomes seamless and

Figure 9. Evaluation Rubric for Student Projects

Evaluation Rubric: Online Lesson (20 points maximum)				
Total Score _____	**(4)** *Excellent*	**(3)** *Good Minor Revisions Needed*	**(2)** *Average Major Revisions Needed*	**(1)** *Poor*
Instructional objectives were met with the lesson design and delivery techniques.	O	O	O	O
Lesson included an icebreaker/opener and/or exercises to gain attention and motivate the learner.	O	O	O	O
Interactive strategies were employed and facilitated an active learning environment.	O	O	O	O
Instructional graphics and handouts enhanced understanding and were easily accessible.	O	O	O	O
Evaluation techniques matched instructional objectives and could be collected in a DE format.	O	O	O	O
Comments:				

transparent.   That also means that we must train our learners to use the technology interface so that it will not become a barrier to learning.

Include an orientation meeting or online tutorial to introduce the technology tools that will be used in your course or program. For example, each student could be sent an e-mail with information about accessing the course and a link to an online orientation.   Once the student has entered the course, the orientation could include an introduction and brief biography of the instructors and/or the distance learning coordinator, links to ordering the textbook(s), links to download plug-ins needed to view the course materials, and online tutorials to introduce the technology tools. It is difficult to separate technology skills from content delivery in mediated-delivered courses. The learner must take initiative in learning the communication technologies necessary for the transmission of the course.

Some instructors choose to have an on-campus orientation. If your learners are close enough in proximity, this is a good way to build rapport. If the course tool you choose is user-friendly, it should include a tutorial on the use of online instructional tools.

In *Advanced Methods*, students demonstrate the use of online tools by submitting a biography to the instructor by WebCT® e-mail and to their virtual team using the WebCT® Threaded Discussion Tool. Edelstein and Edward's (2002) five criteria for evaluating online discussion postings include: (1) promptness and initiative, (2) delivery of post, (3) expression within the post, (4) relevance of post, and (5) contribution to the learning community. Students are also asked to submit the self-assessment using file-saving protocols and the WebCT® Class Assignment Submission feature. They also watch a welcome video and print a PDF copy of the course syllabus to ensure students have the appropriate plug-ins for accessing course materials. In addition, we provide a link for students to order textbooks and to find out who is on their cooperative learning team. This is all done prior to the start of class to ensure that students are comfortable with the format and technology before they delve into the course content (Figure 10).

Provide Information about Appropriate Infrastructure for Learning Support

Effective course design can only go so far. Students must also have access to resources such as libraries and laboratories. This includes providing access to those learners with disabilities. Instruction should seek to help expose and repair barriers to accessibility (Bobby, 2003). Admission and registration materials must be user-friendly and client-driven. This component deserves attention and requires strategic visioning and administrative support from the institution

Figure 10. Advanced Methods Interface Showing Course Orientation

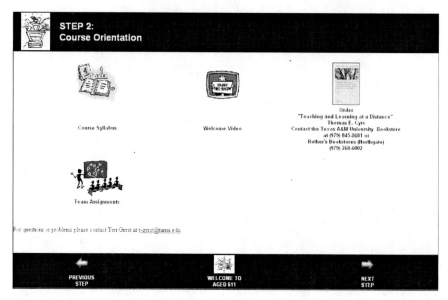

delivering the courses or programs. If you make sure students are aware of and know how to access campus support resources, you will save a great deal of time throughout the duration of a course (Figure 11).

In Figure 1, you may have noticed a tree next to the house. This tree includes links for "branching out" to support services available through the university. Even though these links are available off the main university Web page, students usually go to the faculty member first with questions. Our tree facilitates access to support resources within our online course.

Future Trends

The demands of technology and design are reaching a growing number of programs to provide teams of professionals with the tools to collaborate on the design and delivery of distance learning courses. These teams are driven by the content expert (usually the instructor) who may be assisted by specialists in multimedia, instructional design, and technical communications. Throughout the delivery of a course, the team assists the instructor with technical issues that arise, such as updating course materials on the Web, fielding student questions, and setting up and maintaining equipment. This frees up the faculty member's

Figure 11.   Example of Support Services for Distance Learners

time to focus on his or her students. If you do not have the luxury of an instructional design team, you may choose to work with technically competent students (undergraduate or graduate), who may either earn course credit or hold student assistant positions.

One trend we see continuing to diffuse into distance education is the use of hybrid technologies. Web-based courses began as primarily electronically delivered correspondence courses. They have evolved to include collaborative tools, H.323 Internet-based videoconferencing, and streaming video/audio. With all these multimedia tools available, the concepts discussed in this chapter provide a reliable foundation for developing instructionally effective online learning environments.

To truly make a difference in online instruction worldwide, all educational providers need to work collaboratively to share the best courses and programs to reach a global audience. This brings up many issues that must be addressed for the future, such as copyright and intellectual property, cultural and language-appropriate content, consideration of bandwidth and time zones across the world, and policies and relationships with other universities and industries worldwide.

Conclusion

When designing an online course, keep in mind the following design principles: design or select a course management tool; plan the organization of the course; chunk content into modules; include interactive teaching and learning strategies; incorporate adult learning principles; use self-directed and student-centered learning approaches; use authentic assessment strategies; include an online orientation and technology training; and provide information about the institution's infrastructure for learning support. As our experience in course design grows, we have encountered more online course design challenges than those presented in this chapter. For example, Hall, Watkins and Eller (2003) provide an additional model for Web-based design. We have found, however, that our nine principles serve as a foundation for ensuring that courses provide the best learning experiences possible across programs. We hope you can use these principles to overcome the challenges of creating student-centered learning environments to better serve your students. As we learn, collectively, to be better instructors, we need to continually revise our instructional delivery techniques on the basis of the best science and art available. We hope the material presented in this chapter contributes to both.

References

Bobby. (2003). *Welcome to bobby online service*. Retrieved March 7, 2003 from the World Wide Web: http://bobby.watchfire.com/bobby/html/en/index.jsp.

Bonk, C. J. & Dennon, V. (2003). Frameworks for research, design, benchmarks, training, and pedagogy in web-based distance education. In M. G. Moore and W. G. Anderson (Eds.), *Handbook of Distance Education* (pp. 331-348). Mahwah, NJ: Lawrence Erlbaum Associates.

Buford, J. A. and Lindner, J. R. (2002). *Orientation, Training and Development. Human Resource Management in Local Government: Concepts and Applications for HRM Students and Practitioners.* Cincinnati, OH: South-Western.

Cyrs, T. (1997). *Teaching at a Distance with Merging Technologies: An Instructional Systems Approach.* Las Cruces, NM: Center for Educational Development, New Mexico State University.

Dooley, K. E. and Lindner, J. R. (2002). Competency-based behavioral anchors as authentication tools to document distance education competencies. *Journal of Agricultural Education* 43(1), 24-35.

Dooley, K. E., Lindner, J. R. and Richards, L. J. (2003). A comparison of distance education competencies delivered synchronously and asynchronously. *Journal of Agricultural Education,* 44(1), 84-94.

Edelstein, S. and Edwards, J. (2002). If you build it, they will come: Building learning communities through threaded discussion. *Journal of Online Distance Education Administration,* 5(1). Retrieved March 7, 2003 from the World Wide Web: http://www.westga.edu/%7Edistance/ojdla/spring51/ edelstein51.html.

Eitington, J. E. (1984). *The Winning Trainer.* Houston, TX: Gulf Publishing.

Fosnot, C. T. (Ed.) (1996). *Constructivism: Theory, Perspective, and Practice.* New York: Teachers College Press.

Future is Online. (n.d.) Retrieved January 28, 2003 from the World Wide Web: http://www.mouthshut.com/readreview/24768-1.html.

Grow, G. (1991). Teaching learners to be self-directed. *Adult Education Quarterly,* 41(3), 125-149.

Gunawardena, C. N. and Zittle, F. J. (1997). Social presence as a predictor of satisfaction within a computer-mediated conferencing environment. *The American Journal of Distance Education,* 11(3), 8-26.

Hall, R. H., Watkins, S. E. and Eller, V. M. (2003). A model of Web-based design for learning. In M. G. Moore and W. G. Anderson (Eds.), *Handbook of Distance Education.* Mahwah, NJ: Lawrence Erlbaum Associates.

Hannafin, M. J. (1989). Inter-action strategies and emerging instructional technologies: Psychological perspectives. *Canadian Journal of Educational Communication, 18,* 167-179.

Huba, M. E. and Freed, J. E. (2000). *Learner-Centered Assessment on College Campuses: Shifting the Focus from Teaching to Learning.* Boston, MA: Allyn and Bacon.

Knowles, M. S., Holton, E. F. and Swanson, R. A. (1998). *The Adult Learner.* Houston, TX: Gulf Publishing.

Lindner, J. R. and Dooley, K. E. (2002). Agricultural education competencies and progress towards a doctoral degree. *Journal of Agricultural Education,* 43(1), 57-68.

Lindner, J. R. and Murphy, T. H. (2001). Student perceptions of WebCT in a web supported instructional environment: Distance education technologies for the classroom. *Journal of Applied Communications,* 85(4), 36-47.

Lindner, J. R., Dooley, K. E. and Murphy, T. H. (2001). Discrepancies in competencies between doctoral students on-campus and at a distance. *American Journal of Distance Education,* 15(2), 25-40.

Moore, M. G. (1989). Editorial: Three types of interaction. *The American Journal of Distance Education,* 3(2), 1-6.

Purdy, L. N. and Wright, S. J. (1992). Teaching in distance education: A faculty perspective. *The American Journal of Distance Education,* 6(3), 2-4.

Shearer, R. (2003). Instructional design in distance education: An overview. In M. G. Moore and W. G. Anderson (Eds.), *Handbook of Distance Education.* Mahwah, NJ: Lawrence Erlbaum Associates.

Thach, E. C. and Murphy, K. L. (1995). Competencies for distance education professionals. *Educational Technology Research and Development,* 43(1), 57-79.

Williams, P. E. (2003). Roles and competencies for distance education programs in higher education institutions. *The American Journal of Distance Education,* 17(1), 45-57.

Chapter VI

Theory and Practice for Distance Education: A Heuristic Model for the Virtual Classroom

Charles E. Beck
University of Colorado at Colorado Springs, USA

Gary R. Schornack
University of Colorado at Denver, USA

Abstract

A new world of distance education demands new thinking. Key components to completing the distance educational system requires that institutions determine how the process is designed, delivered, integrated, and supported. Unfortunately, educational administrators tend to view distance education merely as a process of taking existing readings, exercises, handouts, and posting them to the Web. While this approach may seem cost effective, such an approach is not educationally effective. Although the meaningful transition to e-education has just begun, determining measures of effectiveness and efficiency requires innovations in social and political thought beyond the advances in technology. The educational process requires feedback from the professor, from the student, and from the wider community, especially businesses who hire the graduates. As e-learning and higher education reach new heights, they are changing the functions of the university. E-learning changes all the ground rules, including time, distance, and pedagogy. We now have new ways to reach and interact with students, present rich content in courses, and deliver the technologies of the smart classroom to students, wherever they are in the world.

Introduction

Education is now the second largest civilian industry in the US after health care (Dunn, 2001). Distance education is a growth industry in the modern economy, with American's spending over one-half of one trillion dollars on it annually and with over two million classes taken by online education (Shea and Boser, 2001).

As a rapid growth industry, distance education provides a method for both educators and businesses to adjust to new market conditions. Implementing such programs may profit from a systems model for viewing all elements of the educational system. Our approach adapts systems theory to distance education: the systems-based Educational Process Model serves as a heuristic to examine recent research for insight into the distance education process. Using a value-added approach, we are applying the model categories to organize key practices identified from the research. Following the model's categories, we will prepare a list of best practices to help practitioners. Our discussion begins with an overview of the Educational Process Model. With this systems view, we then examine inputs into the system, including the objective educational resources and the subjective philosophy of education. The integration of the model includes purpose (objectives and audience), method (technology and methodology), and pedagogy. The outputs include the objective educational experience, itself, and the subjective outcomes. Assessment provides feedback to the system.

We presented our preliminary ideas at a state-wide conference in Teaching with Technology held in Boulder, Colorado, in June 2002, and at the international Conference of the World Association for Case Method Research (WACRA) in Mannheim, Germany, in July 2002. This chapter represents our most recent research into the new paradigm.

Background

To succeed in modern economies, businesses must help employees acquire updated knowledge and skills, along with the ability to become lifelong learners ("How," 1999). The US Department of Labor projects that 75% of the workforce will need retraining by 2005 (Bersch, 1999). Employee advancement requires the ability to master skills and constantly learn (Messmer, 1999). In addition to employers needing skilled workers, distance education, itself, is a growth industry. In fact, one of the biggest providers of e-learning is the Army's virtual university for enlisted soldiers, which offered online college courses to more than 12,000 students in 2001 (Charp, 2001). According to an estimate in *Forbes*, the online market will become "a $10 billion virtual higher-ed market by 2003 and an $11 billion corporate-learning market by the same year" (Svetcov, 2000). Higher education estimates indicates an increase of distance courses from 5% of students in 1999 to 15% of students by 2002 (Schofield, 1999). Along with the increasing popularity of online learning on the college level, "corporate America is using, distance learning, both internally and externally, for all aspects of training" (Palmer, 2002). According to estimates, corporate training will grow from $2.2 billion to $18.5 billion by 2005 (Charp, 2001).

This new industry is revolutionizing higher education. "Whether it's called distance education, asynchronous studies, online instruction or e-learning, Internet-based training has gained a significant foothold in the realm of professional education. In addition to the time and geographic flexibility, its modalities can be changed to fit the need of the user" (Drew, 2000). Since the time of Socrates, teachers have met students face-to-face for discussion or lectures ("How," 1999). Distance learning changes that basic paradigm. Distance education has evolved from text-based correspondence courses, to videotape-based instruction, to compressed video transmission that allows for two-way audio and video connections between the teacher and the learner (Poole, 2000). In McLuhan's assessment, computer-based training seemed inadequate, a mere "electronic page-turning exercise." However, new developments force a major rethinking (McLuhan, 1998). Up until now, the educational system has remained teacher-centric; but, it is now becoming classroom- and technology-centric. Ultimately, education needs to become learner-centric, using the teachers and technology to unleash students' natural desire for knowledge (Strauss, 2002).

The rapid growth of e-learning has taken many institutions by surprise (Carlson and Repman, 2002). Scholars and practitioners have begun to address this "rethinking" of the distance education process. The advent of digital technologies is now transforming higher education's culture and content. Furthermore, the "technology of higher education is becoming as much a function of market mechanisms as digital media" (Schrage, 2001). As with any new approach, however, skepticism remains. A recent study by a career information and research firm suggests that employers may be cautious about hiring job applicants with "dot-com diplomas" (Jones, 2001). Such concerns recognize that overall, online learning is still in its infancy (Kuchment et al., 2002). A systems view of the field provides one way to raise the level of professionalism in distance education and the Educational Process Model serves to fill that need.

The Educational Process Model

The foundation of the Educational Process Model is a basis system's model which consists of inputs, an integrative process, outputs, and feedback. In adapting this mechanical system to human communication, Beck developed the Rhetorical Process Model, which divided the entire process horizontally into subjective and objective domains (1999, p. 42). The objective process of input-integration-output became status, method, and product. The subjective process became assumptions, purpose, and interpretation. In addition, the integration elements of purpose and method were further divided: purpose includes intentions and audience, while method includes genre and process. These four integration elements represent interactive, rather than linear processes, so embodiment fills the center of the process integrating these four elements. The form of this Rhetorical Process Model serves as the basis for the Educational Process Model, as seen in Figure 1.

Inputs to the Process

The inputs to the educational process include the objective element of resources and the subjective element of educational philosophy. Institutions engaging in distance education must rethink the physical and training needs of the new enterprise and must help faculty expand their philosophy of education in approaching distance learning.

Resources

The obvious resources for distance education involve technology: web sites with sufficient capacity; e-mail, bulletin boards, chatrooms; audio-video streaming;

Figure 1. Educational Process Model

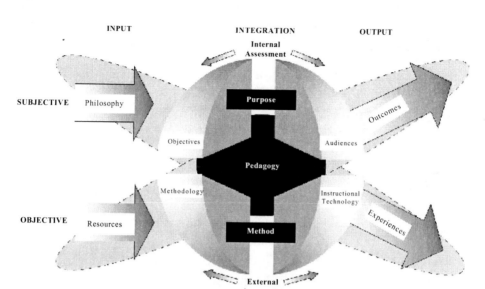

and interactive text, to name a few. The institution also needs technicians who keep the equipment running. Additionally, resources include library access and creating challenges, such as copyright permission and network security (Chepesiuk and Gorman, 1998). Distance education may provide a means for institutions to expand with limited resources. Since the University of Colorado at Denver has one of the highest rates of classroom use, distance education provides a means to expand student capacity without on-campus space (Bethoney, 1997). However, overall costs are driving higher ed both to increase revenues and to become more cost efficient (Heerema and Rodgers, 2001).

More significant, however, are the instructor resources: institutions need to recognize the training and workload implications of distance education. Developing an online course is far more complicated than simply posting a professor's lecture notes online (Green, 2000). According to some estimates, it takes 200 hours to produce one hour of online instruction, thus the need for huge investment, with significant investments of time and money both to develop and revise courses (O'Neal Roach, 2001). "Potential for large-scale usage has been oversold, and return on investment is over-stressed" (Charp, 2001). A seminal article on faculty workload identifies key resource issues: "Is an online course comparable to an onsite campus course in regard to an instructor's teaching load? Even though e-mail and other online requirements may place a greater time demand on the instructor, will the institution recognize this factor? Is there a cap on the maximum number of students accepted into a course? Will an instructor be paid for developing an online course?" (Mirabito, 1996). Rather than merely a theoretical concern, these workload issues reflect significant areas where institutions must adapt to the human resource implication of distance learning. For most institutions, promotion and evaluation committees "may not take technology work seriously" (Young, 2002).

The educational institution must examine these implications and deal with them, recognizing that distance education represents a major shift in instructional resources and faculty development. As institutions move into distance education, they tend to remain faculty centered, following a traditional "one size fits all" model of pedagogy (Johnstone, 2002). Even if the institution attempts to focus on student differences, it largely ignores individual differences among faculty (Gilbert, 2002). In approaching instruction, institutions need to select instructors who have "an eye and ear for how classroom content will translate onscreen." Such faculty must be "receptive to the idea of repackaging in an online format the same content that they have taught in the classroom, in order to take advantage of the additional functionality that the Web allows" (Samuels, 2001).

Ultimately, the move into distance education highlights a major lack in higher education: People who do the majority of teaching at colleges and universities (professors, lecturers, TAs, etc.) get no formal training in teaching at all. Institutions assume that "a professor who cannot use a piece of chalk and a blackboard to teach effectively will be able to do better when we give him or her computers, VCRs, DVD players, PowerPoint presentations, video cameras, the Internet, and a smorgasbord of digital media" (Strauss, 2002). Cheryl White identifies the areas of faculty development needed for an effective distance education program. These recommendations appear in Table 1.

Educational Philosophy

To a certain extent, distance education represents a change to the process: it expands the teacher's potential by removing the physical constraint of buildings and distance (Stoll, 2001). The future context for distance education envisions even greater changes, as outlined by a timeline presented in Table 2 from a recent article in *The Futurist*.

To prepare for such changes, educational institutions should focus on "making our learners smarter, [rather] than ... making obsolete classrooms smarter." To create such a reality, faculty must consider improving their teaching skills "at least as vital as improving their skills doing research and publishing papers"

Table 1.   Faculty Development for Distance Education

Comfort and effectiveness with all technology used in the course
Ability to model use of technology
Ability to track student activities in the course
Willingness to be innovative in teaching methods
Willingness to be innovative in use of technology
Willingness to learn while doing
Willingness to work cooperatively with technical support/design staff
Tolerance of change
Ability to commit significant time to the course
Ability to handle a high amount of interaction with students
Being a good facilitator of communication
Being able to write clear, focused messages
Providing clear expectations of student responsibilities in the course
Ability to design discussions to involve the students

- White, 2000

(Strauss, 2002). Although distance education changes the method of interaction, it actually begins by reinforcing the traditional philosophy of effective education. Table 3 presents the American Association for Higher Education's principles for effective undergraduate education (Merisotis and Phipps, 1999). Jossi would add an additional element to this listing: a key component of success for online education is the instructor's "own enthusiasm for the projects" (2000).

In developing effective education at a distance, institutions must recognize alternate ways to create a learning community. Traditional classrooms provide a "built-in" learning community, that may or may not reach its full potential. Distance learning, however, involves creating a virtual learning community. Even though students access a "virtual classroom" from remote locations, they

Table 2.   The Future of Distance Education

2005	80% of US homes have PCs
2005	Virtual reality is used to teach science, art, history, etc.
2008	All government services delivered electronically
2012	Purely electronic companies exist with minimal human involvement
2015	3-D video conferencing
2025	Learning superseded by transparent interface to smart computers

- The Technology Timeline, 2002

Table 3.   Principles for Effective Undergraduate Education

Encourage contacts between students and faculty
Develop reciprocity and cooperation among students
Use active learning techniques
Give prompt feedback
Emphasize time-on-task
Communicate high expectations and
Respect diverse talents and ways of learning.
                                    -Merisotis and Phipps, 1999

need a sense of belonging to a wider learning objective. Russell identifies the elements needed to create an online learning community (see Table 4).

To create such a community in the new age of technology, the faculty member must become a role model by "demonstrating depth of mastery, wisdom, knowledge, skill, character, and enthusiasm for the subject and profession" (Gilbert, 2002).

Table 4.   Online Learning Community Model

Description	An organization that uses technology to mediate between the individual and collective needs of its members to assure access to tools for learning.
Vision of Adult Learning and Development	Accommodates the special social, psychological, and political characteristics of adult learning.
Learning Contexts	Demonstrates elements of the non-formal, informal, and information-based models of learning.
Indicators of Engaged Learning	Provides learning experiences that are transformative, inclusive of life experiences, rewarding, and accommodating of learning differences.
Instructional Model	Interactive and generative, provides opportunity for customizing adult learning — adapts to a number of learning styles.
Purposes/Goals of Learning	To support collective and participatory communication and to meet a diversity of educational and information needs. - Russell, 1999

Educational Integration

With resources and an effective philosophy, the educator now faces the heart of the educational process: preparing and delivering the content. In terms of the Educational Process Model, this Integrative area focuses on integrative pedagogy, comprised of purpose and method.

Purpose

In a major growth industry, the institutional motive for establishing educational enterprises like distance education "is almost always to make money" (Altbach, 2000). However, the main intention should be tied more to the nature of education itself. The purpose of education requires a dual focus — content and audience. Faculty determines learning objectives, but they do so in light of the specific audience, the learners who opt for a distance program.

Objectives

Distance courses need a different organization from that used in a traditional setting. Institutions must clarify some of these distinctions, as outlined by Sjogren and Fay (Sjogren and Fay, 2002):

- What pedagogy is most appropriate for our goals and our faculty?
- What instructional-design support do we need?
- What are the appropriate assessment tools for evaluating online program quality?

To an extent, the organizing pattern must reflect the instructor's thorough knowledge of the material. The instructor needs the expertise to react to questions and to problems that arise. No longer may the instructor survive "one chapter ahead of students"; the instructor must readily evaluate student capacities and anticipate questions" ("Debating," 2000). In terms of organization, distance education "unbundles" traditional faculty roles into various sub-functions (Armstrong, 2000):

- A knowledgeable professor defines the material to be taught.
- Experts in multimedia pedagogy create the structure of the course.
- Technical people implement it.

- Assessment experts evaluate the course's success in enabling students to learn.

Defining the course and creating the learning objectives for distance education is a time-consuming process: "Developing an effective syllabus and teaching style takes time. E-mail is a different medium from the classroom and requires different techniques to be used effectively by an instructor" ("Debating," 2000). Since the structure of the course must consider both the content material and the learner, the "goal should be to use the new technologies to enhance rather than diminish higher education's critical roles" ("Connections," 2001). This redesign process also requires the students to take greater responsibility for their own education. "If the students aren't doing the work, learning is not occurring" (Strauss, 2002). Cini identifies the ways to design a course to meet the needs of the distance learner, as shown in Table 5.

Audiences

The audiences within the Educational Process Model include the individual students, themselves, as well as how the instructor anticipates the students during course preparation. Even more than in traditional classroom education, distance education can provide a means to support personalized learning (Leinsing et al., 1997). Distance learning does not so much improve traditional methods, but provides a workable alternative for students who need a higher level of flexibility and would otherwise be unable to complete a program (Cutshall, 2002). Additionally, focusing on the student audience recognizes that students not only have differing learning styles, but also differing life needs as they enroll in distance education. On a simplistic level, distance courses are ideal for students facing time and physical constraints ("Debating," 2000). But convenience is not the only issue to address. Expansion in distance learning now reaches a different student body: adults who need to brush up on certain skills,

Table 5.   Criteria for Developing an Effective Online Course

- Assignments that require the learner to actively engage with the course work and one another
- Discussions, activities, and projects that entail collaboration between learners
- A format that fits students' schedules, career goals, and learning styles
- Access to faculty, resources, and classmates
- Convenient access to the class anytime and from anywhere
- Cini, 1998

finish their college education, or begin new endeavors" (Branch, 2001). Most online students are full-time workers who want to improve their skills: to train in the latest software, to qualify for professional advancement, or to participate in some form of education that fits their lifestyle (Green, 2000).

The student audience has begun to take charge. Learners are developing their own personal learning contracts, focusing on their own training objectives, rather than the traditional course instructional objectives (Maise, 1999). Successful students in online courses are a select group: they are "motivated, self-directed individuals who have good computer skills or who are comfortable with technology and not afraid to experiment; they … are able to work alone with minimal guidance" (McLaughlin, 2002). Because student interaction online may fluctuate, online learning must include building a sense that people will notice if a student leaves, or if they "don't get sticky with the content" (Maise, 1999).

Although students tend to seek out distance education opportunities, educators must help students clarify their own expectations for taking such a program. Additionally, students need to learn skills to succeed: how to take notes, how to get organized, and how to deal with the universe of data that obscures the information they actually need to understand. Through the Internet, students have access to millions of books and billions of Web pages, but they need to learn how to use them effectively (Strauss, 2002). Table 6 outlines some of the questions students need to ask. These questions have implications both for the student and for the faculty member.

Table 6.   Student Questions for Online Learning

Student's Questions
What are the computer specifications required for this program?
Am I prepared to deal with the technology and time management demands of online education?
How experienced is the university faculty at offering the course or curriculum, both in subject matter and in dealing with an online curriculum?
How willing are faculty to spend time communicating with students?
Does the online course or program facilitate student interaction in any way?
Am I someone who can learn online?   [Short, 2000]

Faculty Effort
Plan for learning across the lifespan
Identify the distinctive characteristics of adult learners
Emphasize the goals of adults
Create adult-situated contexts for learning   [Russell, 1999]

To determine the overall purpose of distance education, institutions must recognize the need for "smart learners, not smart classrooms." By themselves, smart classrooms will not be enough. For the process to succeed, faculty need an environment that reduces teaching-related administrative tasks so they can focus on developing courses and presenting materials (Strauss, 2002).

Method

The focus on method for distance education usually becomes a focus on technology. Following the Educational Process Model, however, method includes both the technology and the methodology that instructors use in developing and presenting course materials.

Media

Media covers the specific technology components for the instruction, the capacity of the hosting system, and the ability of the student's equipment to interface with the host system. Such consideration usually falls under the domain of the Instructional Technology professionals available to work with the faculty. In addition to providing course content, however, these systems must address security issues, especially with test administration (Mirabito, 1996). Conversion to electronic media impacts both the instructor and the student. From the student's view, the distance approach brings new difficulties, like "Navigating a course's options and jargon — various folders, forums, study guides. ... The ideas are great, the people are great, it's the technology that can be overwhelming if you don't have a background in it" (Loftus, 2001). Table 7 identifies key instructional issues related to selection and use of technology.

Learning technologies have revolutionized the traditional classroom and created new possibilities for eLearning, but technology alone is not the point (Grush, 2002). Technology provides the means of accessing and using knowledge. Inaccessible information remains of little value (Info World, Tom Feldon). Using the information in a distance mode requires developing such a level of trust between the instructor and the student that the student can freely voice doubts and can depend on the instructor to provide guidance in how to use the process ("Debating," 2000). Used properly, both the computer and the Internet permit nonlinear learning strategies where students can move between subjects in their own time and order (Armstrong, 2000). Rather than the medium making the difference, according to Strauss, "It's the particular way in which you use it" (2002). An Internet-based course can hold visitors, bring them back, and become a favorite of the users (Maise, 1999). Done properly, distance education can

Table 7.   Media Implications for Distance Education

Identify new technologies
Become competent with a variety of technology approaches
Demonstrate multiple visual tools
Become comfortable using multiple technology tools
Reinforce new technological tools for communicating Information
Be able to sequence ideas in a logical presentation sequence

- Beck and Schornack, 2001

blend powerful new technology: the Internet, intranets, and e-delivered courses — with traditional media such as instructor-led courses, audio- and videotapes, articles, and books. The result yields a powerful, accessible learning information source (Boxer and Johnson, 2002).

From the instructor's point of view, conversion to distance education requires a significant commitment in both preparation and content time. Faculty must become familiar with the technology: "they must have extensive and responsive technical support at all times" (McLaughlin, 2002). As reported in *Forbes*, converting the curriculum to the Internet is not easy; it takes perhaps 20 times the effort as preparing a traditional classroom instruction, according to Eli Noam, professor of economics and finance at Columbia Business School (Svetcov, 2000). Currently, distance education is labor intensive, but with the growing availability of quality materials, distance ed can eliminate the quantity-versus-quality trade-off (Heerema & Rodgers, 2001). Today's state-of-the-art mediated learning software can incorporate a built-in management component that measures each student's work in detail. Beyond holding students accountable for their own learning, this management tool lets instructors diagnose weaknesses so they immediately know when an individual student falls behind (Barker, 2000).

Methodology

Methods serve as the tactics that educators use to present information. Methodology varies by academic discipline, by educational level, and by the professor's personality. To meet the requirements for effective distance education, however, methodology often involves more than one person:

Courses following this approach must be organized differently from the traditional lecture course and generally involve an "unbundling" of usual

Table 8.   Practical Methods for Developing Effective Distance Education

View your role as coach, facilitator and coordinator
Listen to your students
Don't give answers -- ask questions
Allow students to learn by making mistakes
Encourage students to solve their own problem
Encourage students to do things their own way
Provide guidance and instruction at first -- then decrease direction over time
Be open and honest -- share pertinent information with students
Provide learning opportunities
Measure your success through the success of your students

- Adapted from Cini (1998)

faculty roles. ... The resulting course may contain lectures by the professor who defined the course, a multiplicity of experts lecturing on specific points, or lectures by a hired presenter to reinforce the course's concepts, or it is also possible the course may have no "talking heads" at all (Armstrong, 2000).

Despite these significant variations, faculty can benefit from some practical guidelines that can help them develop a methodology effective for online students. Table 8 identifies these practical methods.

Teaching methods in general, however, must adapt to the distance media, where faculty do not have the traditional contact time with students. As a result, faculty need to adapt their methods to the distance media, following the specific strategies identified in Table 9.

Combining teaching method with technological skills can help students reach higher cognitive levels. Such a process requires that a faculty member becomes a "facilitator, collaborator, and guide who makes instruction learner centered" (Notar et al., 2002).

Pedagogy

Within the Educational Process Model, the embodiment center of the integration process is pedagogy. Pedagogy merges objectives, student audience, technology, and methodology to produce the educational experience. For distance

Table 9.   Specific Strategies

> Let students know when and how you prefer to be contacted, including how often you intend to access e-mails.
>
> Create online forums, threaded discussions, and chats.
>
> Use the Internet to access experts and ideas from "beyond the classroom."
>
> Form student study groups with specific intent.
>
> Keep groups small, suggested optimal size no larger than eight.
>
> Assign teams to work together on projects.
>
> Link each student with a mentor or a specialist in the field.
>
> - Syllabus, David G. Brown

education, this pedagogical effort recognizes that students must invest time to learn how to absorb content via computer, in contrast to traditional approaches. On the practical level, students must even decide "whether and how to take notes" and "how to block out appropriate time" (Aldrich, 2000). The educator's merging of purpose and methods comprise the educator's pedagogy. Teachers modify their original objectives in view of student needs and the available technology. Methods sequences suitable for in-class discussion must change based on the asynchronous nature of the distance process. The pedagogy needed for distance education must overcome significant problems: "that most teaching and learning does not occur in classrooms, that teachers and learners have no formal training in teaching or learning, that we have not developed and deployed the tools that teachers and students need for teaching, learning, and administration" (Strauss, 2002). A pedagogical starting point to address the problems begins with the Integrative process outlined in Table 10.

Although distance education starts as a paradigm shift, as it gets integrated into the curriculum it exerts its own influence on the broader educational process. The lines between distance education and traditional classroom learning are blurred. Defining online and offline instruction as two separate (and sometimes warring) entities represents an outmoded paradigm (Baker, 2000). In creating the new paradigm, technology must not take over the process, making classrooms a complex tangle of technical gadgets that are nearly as difficult to operate as Boeing 747s (Strauss, 2002). A guiding principle can help in this integration — as promulgated early in the prior century by Alfred North Whitehead (1929):

The best education is to be found in gathering the utmost information from the simplest apparatus.

Table 10.   Integrating the Educational Experience for Distance Learners

1. Relate all activities, considering them building blocks for successive activities to be addressed.
2. Employ rich learning activities: use ample examples, especially those that exercise the three higher-levels of cognition because they provide an opportunity for ample plausible interpretations.
3. Use pictures, not text, to the extent possible.
4. Embed the data needed to solve problems in the learning context: learning is the retention and transfer of knowledge to new and different situations.
5. Have students provide "story" resolutions before they are exposed to "expert" solutions.
6. Support multiple links among concepts: two principle objectives to cognitive development include long-term acquisition and retention of stable, organized, and extensive bodies of meaningful, generalizable knowledge. The growth in the ability to use this knowledge in the solution of particular problems means that educators must include those problems which, when solved, augment the learner's original store of knowledge.
7. Present knowledge from multiple perspectives: multiple perspectives are the provision of information such that a learner has the opportunity to utilize any of senses of multiple intelligence that have been identified by Howard Gardner (1999).
8. Use active learning techniques.
9. Stimulate the collaborative process by presenting problems so complex that students must work together to solve them.
10. Support continual self-assessment.
11. Provide support at critical junctures to push students past current limitations.
12. Expose students to expert performance.
13. Provide pairs of related stories (vignettes) to learning to establish transfer outside the macrocontext.

-Notar et al., 2002

Educational Outputs

The outputs of distance education combine the actual experiences of conducting the courses and the ultimate learning outcomes that occur.

Experiences

The output of the Educational Process Model is the actual instruction as presented. Institutions monitor such output in terms of course schedules, equipment usage, contact hours, credit hours generated. After all the preparation and agony, the output represents the "here it is" of education — an objectively monitored event or series of events. Regardless of the availability of technology, the teacher must still produce a stimulating educational experience (Shea and Boser, 2001). For distance education, Fabrotta provides a capsule view of the output experience:

The virtual classroom is more than a buzzword. It represents a classroom furnished with the Internet, e-mail, digital cameras, instant messaging and videoconferencing, enabling students to talk to professors, instructors, industry experts and other students from anywhere in the world using audio and visual components (2000).

Outcomes

Virtual education brings both expected and unexpected outcomes for the institution and for the student. For the institution, Internet-based programs demand more faculty preparation time: Ninety percent of those surveyed say distance learning required more time for preparation than the traditional classroom course ("New," 2001). According to Green, deans report a "dirty little secret ... no one is making money" (Green, 2000). If institutions jump into distance education too quickly without planning and proper allocation of resources, the Internet could potentially weaken enrollment in smaller schools that can't compete with more comprehensive programs (Bascom, 2001).

From the student's perspective, distance education attracts by satisfying practical needs: "convenience, flexibility, and access to low-incidence courses" (Patterson, 2000). But new dynamics emerge as well. Web-based delivery does not inhibit the development of a classroom community; rather, it actually contributes to the formation of a cohesive group (Poole, 2000). Interaction expands, as well: while a few students may tend to dominate a classroom discussion so that shy individuals don't stand a chance, in an "online classroom," both the shy and the pushy can easily speak up (Loftus, 2001). Overall, online courses can foster discussion that relates more directly to the course content (Poole, 2000). Because distance education represents a paradigm shift from traditional education, some studies tend to compare and contrast outcomes between distance and traditional classroom-based learning. Merisotis and Phipps (1999) examined the following areas: student outcomes, such as grades and test scores; student attitudes about learning through distance education; and overall student satisfaction toward distance learning. A more in-depth assessment examines the quality of the interactions involved in the distance process. For example, after experience with a leadership program, Cini analyzed the

Table 11.   Outcomes of Online Course Effectiveness

Data Measures
- The quality and quantity of student comments and online discussion
- The number of public messages sent to the discussion group versus the number of private messages sent to the instructor only
- The grades for the course

Findings
- A deeper level of discussion emerged online than in a classroom setting, enhancing critical thinking
- Students participate more actively; shy students are as involved as more outgoing students
- Dialogue between students improved communication skills
- A decreased dependence on the instructor as the sole source of knowledge
- The instructor's role changes to that of facilitator and mentor
- Student access to informational resources is expanded
- Learners become less dependent on the instructor
- Education becomes learner-centered, as opposed to instructor-centered
- Interactions between learners are increased
- Collaboration between instructors and learners is the norm
- Instructors become learners and learners become instructors

- Adapted from Cini, 1998

online comments of students and the direction of flow, whether among students themselves or directed toward the instructor. Table 11 presents these findings.

Assessment

Assessment serves a critical feedback role in the educational process. In terms of the Educational Process Model, assessment provides feedback for the entire system. Such feedback may not occur instantaneously, but only after the institution has gained some experience. Assessment begins internally, but then involves outside agencies, as well.

Internal

Distance education begins with an initial adaptation to technology, to students, and new teaching methods. Ultimately, these changes begin to affect the rest of the institution, thus the systems nature of the Educational Process Model. According to Armstrong, after computer-mediated distance learning becomes comparable in learning effectiveness to traditional lecture courses, colleges and universities will begin to experience pressure to bring similar technologies into teaching within their on-campus programs (Armstrong, 2000). But changing technology brings consequences. Beyond the faculty workload mentioned earlier, institutions must recognize the significance of such efforts for faculty promotion and tenure. As Young indicates in a recent article in the *Chronicle of Higher Education*, "Work creating online teaching materials — such as sophisticated Web sites or multimedia tools designed to help students — slips through the cracks of the three traditional categories used in promotion" (2002). And even institutions which experience early success in distance education have no guarantees. Who, after all, is going to attend Virtual U when they can go to Harvard online instead? (Green, 2000)

Informal and anecdotal evidence about online education finds that, "basic demographic characteristics such as gender and age are not reliable predictors of cyber-student performance," and that students who were members of a cyberstudy group had higher final grades in our class than those who preferred to study alone (Wang and Newlin, 2002). Instructors must become more aware of differences to identify at-risk students, because the usual cues associated with student anxiety, inattentiveness, or apathy are not present in the virtual classroom. Specifically, instructors need to examine low "hit rates" to the course home page and inactivity in writing or reading forum postings (Wang and Newlin, 2002). More formal findings appear in the results of a survey of faculty in distance education, as shown in Table 12.

Table 12. Six Best Outcomes of Online Distance Delivery

Tutorial functions	Learners are able to use the system for study of basic math concepts and review/remediation functions, allowing more time for individual faculty/learner contact and discussion.
Flexibility	Adult learners, who have significant professional and personal time constraints, have the option of working anytime, anywhere.
Self-paced	Learners are not stalled by predetermined course schedules; they may complete assignments in as little or as much time as necessary.
Privacy	Learners can operate in a private environment, interact in a computer-adaptive environment, and efficiently focus on concepts and areas they need.
Cutting edge	Developmental education learners, traditionally offered second-rate services, were being offered an attractive state-of-the-art option.
Interactive feedback	Through a computer-adaptive environment, learners receive immediate constructive feedback after each response rather than having to wait for lesson or test results to monitor progress.
	- Perez and Foshay, 2002

External assessment involves outside experts, government agencies, or education associations examining individual schools or programs. According to one overview, "The 'factory' university is giving way to the 'virtual' university" (Dunn, 2001). This shift has gained the involvement of unions and accrediting agencies. The American Federation of Teachers has recommended quality standards for content support, technical support and counseling for students, protection of intellectual property rights, and proper training for faculty ("New," 2001). The California Virtual Campus presented the first Teaching Site Awards to recognize exemplary online classes in the areas of educational content, course design, use of multimedia, interactivity, community, and disabled access (*Matrix*, 2001). In Fall 2001, *U.S. News & World Report* identified the top 26 distance M.B.A. Programs ("E-Learning," 2001).

While these activities represent traditional methods of institutional assessment, the virtual classroom brings further assessment implications. The virtual university is "not a single institution, but a web of providers which collectively provide educational services at the time, place, pace, and style desired by the client, with quality determined by the client and a variety of approving and accrediting bodies." Such a process will lead to "two main types of educational institutions — those adding value in course work and those that are certifying agencies" (Dunn, 2001). Because distance education requires a rethinking of the process, it can lead to a paradigm shift in educational theory and practice. Ultimately, technology does not represent a minor change in delivery; rather, it is an integral part of the educational system, where changing one aspect affects the rest of the system. It can lead to the major assessment considerations for institutions, as presented in Table 13.

Conclusion

Online learning represents one of the most important developments of the past 100 years of higher education (Learner, 2002). The interest in distance education forces educators to rethink the entire process. To guide this rethinking, we are presenting the Educational Process Model as a way to identify elements of the entire distance process. We have applied this model to selected research findings as the initial phase of our ongoing research. The model's categories identify the elements of the educational process and provide a way to group the findings of existing research. As we look toward the future, some educators see distance education as "the greatest door-opener to knowledge in recorded

Table 13. Issues to Consider in Implementing Online Learning

- Understand why distance learning is used as specific institutions
- Enter into agreements with for-profit entities with great caution
- If new revenue generation is a clear institutional motivation, understand the economics
- If new revenue generation is a clear institutional motivation, watch the quality
- Try for more than automation of the traditional classroom
- Faculty compensation and intellectual property are significant issues; use short-term agreements and understand the economics

Adapted from Berg (2001)

history" (Thomas, 1999). This growing distance industry will have the incentive to create even better technology to improve the process. Technologists are working toward tele-immersion, an "advanced shared environment" that will one day create the illusion that you are in the same room with people at a distance (Virtual Roundtable, 2000). However, we are not yet at that juncture. Both teachers and students need to learn new skills. In particular, they face the "daunting task" of overcoming "FUD — fear, uncertainty, and doubt" (Patterson, 2000).

E-learning and higher education are reaching new heights and are changing the functions of the university. E-learning has changed the ground rules of everything, including time, distance, and pedagogy. We now have new ways to reach and interact with students, present rich content in courses, and deliver the technologies of the smart classroom to students wherever they are in the world. In evaluating the learning process, educators must get feedback from their students and potential students. An exchange in the "Virtual Roundtable" captures this need:

Schaaf: "You've suggested that educators have not traditionally looked for feedback from the learner."

Schank: "That's exactly right, because they don't give a damn about students. But, we do. You have to because people are going to be able to vote finally [take a virtual course at another school]" (2000).

Although educators may see distance education as a "rethinking" or a "paradigm shift," the "Virtual Roundtable" addresses this process in a more succinct fashion. As the second-half of the title states, "the e-Learning Revolution Is Not about Computers, it's about Communication." While we have used a communication-based systems model to examine the process, Mary Furlong makes the communications aspect alive ("Virtual Roundtable," p. 72):

You go online because of information.
You stay online because of the relationships that form.

References

Aldrich, C. (2000). Customer-focused e-learning: The drivers. *Training and Development*, 54(8), 34.

Altbach, P. G. (2000). The crisis in multinational higher education. *Change*, 32(6), 29.

Armstrong, L. (2000). Distance learning: The challenge to conventional higher education. *Change*, 32(6), 20.

Barker, J. (2000). Sophisticated technology offers higher education options. *T H E Journal (Technological Horizons in Education)*, 28(4), 58.

Bascom, L. C. (2001). Budgetary blues: As the annual higher education budget battles heat up, new challenges ranging from the repeal of the estate ax to the Internet pose bottom line problems. *Matrix: The Magazine for Leaders in Higher Education*, 2(3), 22-26.

Beck, C. E. (1999). *Managerial Communication: Bridging Theory and Practice*. Prentice-Hall.

Beck, C. E. and Schornack, G. R. (2001). Electronic communication: The challenge to innovate. *Journal of American Academy of Business, Cambridge*, 12(1), 122-130.

Berg, G. A. (2001). Distance learning best practices debate. *WebNet Journal*, 3(2), 5.

Bersch, C. (1999). Making the grade. *Communication News*, 36(12), 18.

Bethoney, H. (1997). Easing the learning slope. *PC Week*, 14(43), 50.

Boxer, K. and Johnson, B. (2002). How to build an online learning center: Online learning centers may be the new construct for training and development. *T&D*, 56(8), 36.

Branch, A. (2001). Thomson busy, then shuts down Harcourt's higher education web site. *Matrix: The Magazine for Leaders in Higher Education*, 2(4), 13-14.

Brown, D. G. (2002). Connecting with students. *Syllabus*, 16(3), 24.

Carlson, R. and Repman J. (2002). Show me the money? Rewards and incentives for e-learning. *International Journal on E-Learning*, 1(3), 9.

Charp, S. (2001). E-learning. *T H E Journal (Technological Horizons in Education)*, 28(9), 10.

Chepesiuk, R. and Gorman, M. (1998). Internet college: The virtual classroom challenge. *American Libraries*, 29(3), 52-55.

Cini, M. A. (1998). Learning leadership online: A synergy of the medium and the message. *Journal of Leadership Studies*, 5(2), 103.

Connections for a New Millennium. (2001). *Change*, 33(1), 4.

Cutshall, S. (2002). When online learning works. *Techniques*, 77(5), 22.

Debating Distance Learning. (2000). *Communications of the ACM*, 43(2), 11.

Dunn, S. L. (2001). Fuel for the future. *USA Today (Magazine)*, 129, 28.

E-Learning: Best Online Graduate Programs: Business. (2001). *U.S. News and World Report*. Retrieved from the World Wide Web: www.usnews.com/usnews/edu/learning/rankings/mba_reg_prof.htm.

Frabotta, D. (2000). Continuing education: Cornell University's dean-elect plans to focus on research, technology-based learning. *Hotel and Motel Management*, 215(10), 7.

Gilbert, S. W. (2002). Personalizing pedagogy. *Syllabus*, 16(3), 26.

Green, J. (2000). The online education bubble. *The American Prospect*, 11(22), 32.

Grush, M. (2002). Editor's note. *Syllabus*, 16(2), 3.

Heerema, D. L. & Rodgers, R. L. (2001). Avoiding the quality/quantity trade-off in: Distance education. *T H E Journal (Technological Horizons in Education)*, 29(5), 14-19.

How the Web Is Revolutionizing Learning. (1999). *Business Week*, 3661, 40.

Johnstone, S. M. (2002). U.S. distance learning: A 'cottage industry.' *Syllabus*, 15(7), 18.

Jones, C. (2001). Out of a Cracker Jack Box? As online degrees start showing up on people's resumes, what will the business world think of them? *Online Learning*, 5(4), 51.

Jossi, F. (2000). Online adventures: The motivation factor. *Technology & Learning*, 20(9), 32.

Kuchment, A., Binlot, A. and Ferro, C. (2002). Acing Oxford – from home. *Newsweek International*, June 10, 68.

Learner, N. (2002). Experts: New 'paradigm' needed for online learning. *Education Daily*, 35(183), 1.

Leinsing, D., Rosen, L. and November, A. (1997). Building a Community of Excellence. *National Forum*, 77(1), 31-37.

Loftus, M. (2001). But what's it like? *U.S. News & World Report*, October 15, 56.

Masie, E. (1999). Creating stickier online learning sites, classrooms. *Computer Reseller News*, June 7, 56.

_____. (2001). *Matrix: The Magazine for Leaders in Higher Education*, 2(1), 62.

McLaughlin, J. (2002). Getting online. Getting up to speed. *International Journal on E-Learning*, 1(3), 21.

McLaughlin, J. (2002). Surf or turf? Deliberations about distance education. *International Journal on E-Learning*, 1(2), 18.

McLuhan, R. (1998). The online learning curve. *Computer Weekly*, July 2, 38.

Merisotis, J. and Phipps, R. (1999). What's the difference? *Change*, 31(3), 12-17.

Messmer, M. (1999). Skills for a new millennium. *Strategic Finance*, 81(2), 10-11.

Mirabito, M. (1996). Establishing an online education program. *T H E Journal (Technological Horizons in Education)*, 24(1), 57-60.

New AFT Report Proposes Standards of Online Programs. (2001). *Black Issues in Higher Education*, February 1, 43.

Notar, C., Wilson J. and Ross, K. (2002). Distant learning for the development of higher-level cognitive skills. *Education*, 122(4), 642.

O'Neal Roach, J. (2001, June). E-learning: Is it the end of medical schools? *Student BMJ*, 174.

Palmer, P. (2002, July). From a distance: Getting a degree in your time and space. *Black Enterprise*, 52.

Patterson, J. (2000). Building the virtual classroom. *Curriculum Administrator*, 36(7), S14.

Perez, S. and Foshay, R. (2002). Adding up the distance: Can developmental studies work in a distance learning environment? *T H E Journal (Technological Horizons in Education)*, 29(8), 16.

Poole, D. (2000). Student participation in a distance-oriented online course: A case study. *Journal of Research on Computing in Education*, 33(2), 162.

Robb, D. (2000). At home with Internet-based training. *Risk Management*, 47(7), 27.

Russell, M. (1999). Online learning communities: Implications for adult learning. *Adult Learning*, 10(4), 28.

Samuels, D. (2001). On learning online. *Risk Management*, 48(2), 8.

Schofield, J. (1999). Back to school online: For thousands of students, online learning is proving to be a match made in cyberspace. *Maclean's*, September 6, 22.

Schrage, M. (2001). Brave new world for higher education. *Technical Review*, 104(8), 90.

Shea, R. and Boser, U. (2001). So where's the beef? *U.S. News & World Report*, October 15, 44.

Short, N. (2000). Online learning: Ready, set, click. *RN*, 63(11), 28.

Sjogren, J. and Fay, J. (2002). Cost issues in online learning: Using 'co-opetition' to advantage. *Change*, 34(3), 53.

Stoll, J. (2001). No time for class? Log on to Internet. *Automotive News*, 75(5915), 28.

Strauss, H. (2002). New learning spaces: Smart learners, not smart classrooms. *Syllabus*, 16(2), 13.

Svetcov, D. (2000). The virtual classroom vs. the real one. *Forbes*, 166(7), 50.

The technology timeline. (2002). *The Futurist*, 36(5), 33.

Thomas, S. (1999). Adult learning goes online. *Computer Shopper*, 19(10), 331.

A virtual roundtable: Why the e-learning revolution is not about computers: It's about communication. (2000). *Training*, 37(9), 64-76.

Wang, A. and Newlin, M. (2002). Predictors of performance in the virtual classroom: Identifying and helping at-risk cyber-students. *T H E Journal (Technological Horizons in Education)*, 29(10), 21.

White, C. (2000). Learn online. *T H E Journal (Technological Horizons in Education)*, 27(9), 66.

Whitehead, A. (1929). *The Aims of Education and Other Essays.* MacMillan.

Young, J. (2002). Ever so slowly, colleges start to count work with technology in tenure decision. *Chronicle of Higher Education*, February 22, A25.

Chapter VII

Looking for Indicators of Success for Distance Education

Wm. Benjamin Martz, Jr.
University of Colorado at Colorado Springs, USA

Venkateshwar K. Reddy
University of Colorado at Colorado Springs, USA

Karen Sangermano
University of Colorado at Colorado Springs, USA

Abstract

The purpose of this chapter is to identify key components of distance education satisfaction. The distance education environment is an expanding market driven by several market forces. A working list of potential variables for satisfaction can be developed from the previous research done to compare the traditional to the distance education environments. A questionnaire was developed using these variables and administered to 341 distance students in a successful, top 26, M.B.A., distance education program. The results of the questionnaire are factored into five constructs that ultimately correlate well with the satisfaction ratings of the subjects. Using these factors as guidance, some operational and administrative implications of those findings are discussed.

Introduction

The primary market driver for Distance Education is that it is the major growth segment in the education industry. In 1999, nearly 80% of the public, four-year institutions and over 60% of the public, two-year institutions offered distance education courses. More than 1.6 million students are enrolled in distance courses today. Technology advancements, such as the Internet, video streaming, voice-over IP, groupware, intranets, etc., are enabling this vast market. Traditional education providers, such as universities, see the market as a way to expand. More than 90% of all colleges are expected to offer some online courses by 2004 (Institute of Higher Education Policy, 2000). Corporations envision online training warehouses saving large amounts of training dollars. Combined, the virtual education market and its sister market, corporate learning, are predicted to grow to more than $21 billion by the end of 2003 (Svetcov, 2000).

Other Market Drivers

There are several other drivers in the growth and acceptance of distance education. First, job market expectations are changing. In the past, employees and employers maintained employment relationships over long periods of time. Today, employees are not expected to stay in the same job for long periods of time. The current modes of careers include: multiple careers, combinations of part-time work in multiple jobs, telecommuting, leaving and then reentering the full-time work force, switching jobs, etc. Today's employee easily envisions the need to maintain a level of knowledge current with his or her career demands (Boyatzis and Kram, 1999). The concept of lifelong learning has emerged. Lifelong learning is the idea that people, in order to respond to the quickly changing work environment, will need to perform continuous learning throughout their work careers.

The educational institution, itself, sees a growth opportunity. The program implemented must make operational and financial sense, but may be seen as a way to make money (Shepherd et al., 2002). For example, elective classes that do not have sufficient enrollments on-campus may pick up enough distance students to make teaching the course more feasible (Creahan and Hoge, 1998). Costs savings may be obtained and, if significant enough, may drive up demand as costs may be lowered. Finally, most educational institutions serve a geographical region, either by charter or by mission and a distance learning program may be a practical method to help satisfy this strategic mission (Creahan and Hoge, 1998). There can be problems, too. For example, the administrative premise that the distance education program will save money (Creahan and Hoge, 1998) can be both problematic and counterproductive. This may not prove true as the technological infrastructure (i.e., network servers, student support), along with the faculty training required for an above-average distance program, is substantial. Finally, if successful, the complexity caused by the sheer volume of large distance classes may not be anticipated.

The sheer size and complexity of this opportunity significantly raises the stakes for those organizations undertaking distance education. Customer satisfaction is one of the key factors contributing to the Malcolm Baldridge awards, a top award for quality in the commercial business world. In the new Malcolm Baldridge evaluation criteria, (ISO9001: 2000), companies are asked to better show a program's effectiveness through customer satisfaction. Vavra (2002) proposes that the best way to demonstrate effectiveness is to "correlate customer satisfaction with customer behavior." Gustafsson et al. (2000) show customer satisfaction linked significantly to quality at the Volvo Car Corporation. In their more broad analysis of well-run companies, Peters and Waterman (1982) deemed customer satisfaction as a key factor contributing to the companies' performance.

With this perspective in mind, it makes sense that satisfaction, as an important measure of quality, transfers to distance education programs. One of the key factors to a program's success will be the satisfaction of one of its key stakeholder — its students.

Impact of Commercialization

However, the "commercialization" of education raises its own concerns about the basic process of learning (Noble, 1999). For example, are there any problems fundamental to the distance environment because of limited social interaction? Retention may be one such problem. Carr (2000) reports a 50% dropout rate for online courses.

Haythornthwaite et al. (2000) think they've found another one. They coined a term, "fade back," to describe when students do not participate in the distance class. They looked at how social cues, such as text without voice, voice without body language, class attendance without seating arrangements, and students signing in without attending Internet class impacted students "fading back." They found that the likelihood of students "fading back" is greater in distance learning classes than in face-to-face classes.

Hogan and Kwiatkowski (1998) report on the social aspects of teaching in large groups in the United Kingdom. Their argument includes the premise that technology can handle the activity of teaching to large groups; However, the emotional aspects of this teaching method have been ignored. Similar concerns are raised from Australia, where technology has been supporting distance teaching for many years. Hearn and Scott (1998) argue that before adopting technology for distance teaching, that technology must acknowledge the social context of learning.

Historically, researchers have been exploring the impact of technology on education since 1928 (Russell, 1999). These technologies have included correspondence classes, broadcast and recorded video, audio records and tapes, audio-graphic, video and computer conferencing (Moore, 1995). For the purposes of this discussion, distance education is defined as "creating a learning environment that facilitates structured learning without the traditional practice of face-to-face interaction in an on-campus environment" (Martz and Shepherd, 2002).

These studies raised other areas of concern. Tinto (1975) compared the learning retention of distance groups with traditional groups and found that social integration was a key factor in successful retention of traditional groups. Dixon's (2001) feedback studies found that timely feedback was a key to the self-worth and satisfaction with distance courses reported by students. Several researchers and educators are studying how distance education classes create a sense of community (Haythornthwaite et al., 2000; Dede, 1996). Finally, and somewhat intuitively, computer skills (Carabajal, 2000) have been found to positively impact student satisfaction with distance classes.

Two other factors, trust and isolation have been suggested by Kirkman et al. (2002). Studies by Ragan (1998) and Kirkman et al. (2002) showed that "frequent interaction" and "rapid responses," respectively, helped foster trust in the virtual environment. Kirkman et al. (2002) went further and suggested that "establishing norms" for communication patterns is a key factor in trust development. Finally, Kirkman et al. (2002) identified the feelings of isolation as a major issue for virtual environments such as distance education.

Changes to the Educational Environment

By definition, the paradigm of distance education changes the education environment. This means that students will probably respond differently to this environment than they do to the traditional classroom. For example, Palloff and Pratt (1999) conjecture that those students considered introverts will handle the electronic environment better than their extroverted counterparts. Their logic suggests that the distance environment's characteristic of a delayed, thoughtful process will fit better with the basic interaction activity of an introvert than with an extrovert. Further, as Brookfield (1995) notes, this change impacts the faculty, as well. For popular, personable faculty, this distance environment will minimize this positive attribute.

Distance education is also impacting the basic pedagogy of education. Bilimoria and Wheeler (1995) posit a model called Learning Centered Education (LCE), which "places learning and learners at the core of the educational process." Essentially, a learning partnership is created whereby the teacher identifies what needs to be learned and the student helps identify the means by which their own learning occurs. The student helps to customize his or her own learning. The concept works because it sets clear expectations on both sides of the relationship. The teacher facilitates and the student participates.

Finally, academic researchers have always been interested in explaining how people react to the introduction of technology. This is true of the distance education environment. With regard to how people learn, Polanyi (1966) suggested a recurring cycle of today's new technology becoming tomorrow's commonplace technology. Poole and DeSanctis (1990) suggested a model called Adaptive Structuration Theory (AST). The fundamental premise of the model is that the technology under study is the limiting factor or the constraint for communication. It further proposes that the users of the technology, the senders and the receivers, figure out alternative ways to send information over the channel (technology). The term used for this adaptive activity is "appropriation" and it "refers to the manner in which structures are adapted by a group for its own use ..." (Gopal et al., 1993). A good example here is how a sender of e-mail may use combinations of keyboard characters or emoticons (i.e., :) — sarcastic smile; ;) — wink ; :o — exclamation of surprise) to communicate more about their emotion on a subject to the receiver. For distance education, this model would imply more satisfaction: (1) if the technology used establishes an adequate communication channel, or (2) if the technology is used over time.

Despite the large number of students, educators and organizations involved with distance education, the debate still rages as to its effectiveness. The complex environment contains issues around how students learn, what motivates students to learn, how to make the learning environment better, how to make the environment worthwhile for all the stakeholders and how technology impacts the learning environment. The study of these complex issues and interactions falls into the broad field of social informatics that Kling (2000) defines as the study of the interaction of information technologies with an institutional and cultural perspective.

Is distance education comparable to the traditional classroom? The simple answer is: maybe. A review of more than 400 studies compared distance to traditional classroom instruction with complex and conflicting results (Russell, 2001). On the positive side, faculty at eCollege.com reported that their students learned equally effective online as they did on campus (TeleEducation NB, 1999). GSS studies are identifying and reconfirming the important characteristics in developing a sense of community in virtual teams, such as satisfaction (Chidambaram and Bostrom, 1997); trust (Aranda et al., 1998); cohesiveness

(Chidambaram, 1996) and participation (Nunamaker et al., 1991). However, a distance learning environment brings with it additional concerns, such as higher rates of student anxiety and frustration (Hara and Kling, forthcoming). In total, though, the preponderance of studies currently reported found "no significant difference" in the learning process between traditional in-class and distance modes.

While the results of all of these conjectures, hypotheses and studies remain inconclusive in settling the distance-versus-traditional-educational-environment debate, they do provide insights for studying satisfaction in the distance environment. The fundamental concerns for differences in the two environments create a great working set of variables for our purposes here. We can adopt our predecessors' rationale as we look for indicators of satisfaction within distance education. A short list of potential indicators includes: computer skills, effective use of technology, fairness, interaction and feedback, technology support, program reputation, the characteristics of traditional classroom (multiple communication channels, personal and social interaction), isolation, workload, customized learning, course content, learning and satisfaction.

The Research Study

Methodology

The program studied is one of the largest, online, AACSB-accredited M.B.A. programs in the world (*U.S. News & World Report*, 2001). Authors had access to proprietary information, such as grades and class e-mail lists, allowing for a richer analysis of the data. Together, these characteristics increase the overall credibility of the results for the audience.

The methodology used is an exploratory factor analysis (Tucker and MacCallum, 1997) in nature. Sample surveys such as this fit well with McGrath's (1984) discussion of research techniques. His discussion of good research design focuses on addressing three criteria: generalizabilty; context; and precision of measurement of desired behavior. McGrath concedes that no research methodology can maximize all three criteria, but suggests that a good methodology will consider the impact of all three. As a sample survey of distance education participants, our methodology maximizes the generalizability criteria (McGrath, 1984). The other two criteria are addressed secondarily in the design, as the subjects are still in the environment with which the research is concerned — a distance education program. This means that the subjects are still in the appropriate context and that the ratings reflect current behavior. As such, the context and measurement of desired behavior confounds should be somewhat minimized.

Instrument

A questionnaire with a battery of 49 questions was developed using the literature discussed earlier as a guide. First, each subject was asked to identify a reference course about which to answer the questions. Once the subject identified his or her reference course, that subject's grade was obtained from administrative records and recorded. In addition, four other demographic questions gathered information on gender, number of courses taken, student status, amount of time expected to spend in the reference course, and the amount of time actually spent in the reference course.

Two sets of five-point Likert (1 = Strongly Agree, ... 5 = Strongly Disagree) questions were used. The first set of nine questions concentrated on the technology used in the reference course. Each two-part question asked if a particular technology (e-mail, electronic forums as a group or class, online tests, group projects, electronic announcements, phone calls, electronic gradebook, fax/postal mail) was used and, if so, how effective was its usage? An additional data item, LOHITECH, was developed by placing the subjects into two groups. One group reported using three or less technologies, while the other group reported using four or more technologies in their reference class. This classification was created to allow an analysis correlating satisfaction to the level of technology used.

The second set of questions asked students to rate their experience with the reference distance course against statements concerning potential influences for satisfaction. These questions associated a five-point rating scale to statements about the impact of technology, course content, tests, instructor interaction and feedback, critical thinking, Association to Advance Collegiate Schools of Business (AACSB) accreditation, workload expectations, socialization and satisfaction. The order of the questions was randomly determined and the questionnaire was reviewed for biased or misleading questions by non-authors.

An example of the questionnaire is provided in Appendix A. Other students and researchers previewed the questionnaire before it was sent to the subjects. The questions are labeled for reference purposes.

Subjects

The questionnaire was sent to 341 students enrolled in the distance M.B.A. program. In Fall 2002, the program served 206 students from 39 states and 12 countries. The majority of these students are employed full-time. The program used in this study has been running since Fall 1996 and has over 179 graduates. It offers an AACSB accredited M.B.A. and its curriculum parallels the on-campus curriculum. Along with the general M.B.A., there are seven different areas of emphasis offered to customize the M.B.A. for a student's interest. Close to 33% of the enrolled students are female. The oldest student enrolled is 60 years old and the youngest is 22. The average age of all students enrolled is 35. More than 25 Ph.D. qualified instructors participate in developing and delivering the distance program annually. Recently, the news magazine *U.S. News & World Report* (2001) classified the program as one of the top 26 distance education programs.

There were 131 usable questionnaires returned. The student's final grade for their reference course was obtained and added to the questionnaire record as a variable. These were separated into two groups: 30 that had not yet taken a course and 101 that had completed at least one course. This second group, those students who had completed at least one course, provided the focus for this study.

Results and Discussion

Question 24, "Overall, I was satisfied with the course," was used as the subject's level of general satisfaction. The data set was loaded into SPSS for analysis. Table 1 shows that 23 variables, including LOHITECH, proved significantly correlated to satisfaction (Q24).

The large number of significant variables leads to the need for a more detailed analysis on how to group them (StatSoft, 2002). Kerlinger (1986) suggests the use of factor analysis in this case "to explore variable areas in order to identify the factors presumably underlying the variables." Kim and Mueller (1986) further suggest that a factor analysis can be used for data reduction and to obtain factors that can be used for further study. A SPSS factor analysis was performed using Principal Component Analysis with a Varimax Extraction on those questions that had proven significantly correlated to satisfaction. Table 2 shows the reliability factor (Cronbach's Alpha) and eigenvalue for each of the five components derived from the factor analysis.

Four questions, Q28, Q39, Q23, and Q37 offered somewhat ambiguous loading factors on the components. Ultimately, each of these was combined into a factor based upon: (1) the addition of the variable to the factor did not reduce the component's reliability below .7000, and (2) the topic of the question entered matched with at least one of the original variables and theme of the component.

Table 1.   Questions that Correlate Significantly to Satisfaction

ID	Question Statement	Correlation Coef.	Sign.
16	I was satisfied with the content of the course.	.605	.000
17	The tests were fair assessments of my knowledge.	.473	.000
18	I would take another distance course with this professor.	.755	.000
19	I would take another distance course.	.398	.000
20	The course workload was fair.	.467	.000
21	The amount of interaction with the professor and other students was what I expected.	.710	.000
22	The course used groups to help with learning.	.495	.000
23	I would like to have had more interaction with the professor.	-.508	.000
26	The course content was valuable to me personally.	.439	.000
28	Grading was fair.	.735	.000
30	Often I felt "lost" in the distance class.	-.394	.000
31	The class instructions were explicit.	.452	.000
33	Feedback from the instructor was timely	.592	.000
34	I received personalized feedback from the instructor.	.499	.000
36	I would have learned more if I had taken this class on-campus (as opposed to online).	-.400	.000
37	This course made me think critically about the issues covered.	.423	.000
38	I think technology (e-mail, Web, discussion forums) was utilized effectively in this class.	.559	.000
39	I felt that I could customize my learning more in the distance format.	.254	.001
42	The course content was valuable to me professionally.	.442	.000
43	I missed the interaction of a "live," traditional classroom.	-.341	.002
46	Overall, the program is a good value (quality/cost).	.258	.017
LOHITECH	Aggregate of Yes votes in Q6 through Q15	.270	.012

Consequently, all reliability coefficients (Cronbach Alpha) are above .7000, indicating an acceptable level for a viable factor (Kline, 1993; Nunnally, 1978). Finally, the five components explain 66.932% of the variance. We discuss possible interpretations of the factors next.

Interaction With Professor: The first component for study combines five questions: Q18, Q21, Q28, Q33, Q34. The themes permeating these questions seem to center on feedback and interaction with the professor. The feedback statements Q33 and Q34 show that feedback that is timely and personalized helped raise the satisfaction ratings. The other two, Q21 and Q28, show the relationship with the professor to be important. Q21 asks explicitly about the level of interaction with the professor, while Q18 gets at the same thing implicitly — since a student reporting high satisfaction is more likely to take another course with the same professor.

Fairness: Q17, Q20, Q28 and Q31 create the second construct. Clearly, Q17, Q18 and Q20 fit this description. All three ask explicitly about the fairness of the tests, the course workload and grading. Q31 asks about the explicitness of the instructions given for the course. This variable fits into the construct by realizing that more explicit instructions are probably setting expectations for the course.

Table 2.   Result of Factor Analysis

ID	Component and Loading Factors				
	1	2	3	4	5
Q18	.576				
Q21	.643				
Q33	.794				
Q34	.848				
Q17		.722			
Q20		.738			
Q28		.411			
Q31		.512			
Q16			.596		
Q26			.850		
Q39			.308		
Q42			.825		
Q23				-.354	
Q30				-.514	
Q36				-.809	
Q43				-.770	
LOHITECH					.508
Q19					.596
Q22					.542
Q37					.494
Q38					.478
Q46					.700
Reliability(1)	.8573	.7715	.7286	.7948	.7358
Eigenvalue	8.520	1.853	1.630	1.541	1.181
(1) SPSS – Cronbach's Alpha					

Courses with changes in instructions or that have ambiguous instructions are probably perceived less fair.

Course Content: The third component derived seems to center around course content. First, three of the questions, Q16, Q26, and Q42, explicitly focus on course content. It seems likely that "good" course content would influence the basic satisfaction of that course. Note that content is perceived from both a personal and professional level. While Q39 about the student's ability to customize learning has a low loading factor, it fits here when we realize that if a student is customizing their learning, they are choosing course content.

Classroom Interaction: The questions composing the fourth construct, Q23, Q30, Q36, Q43, all have negative loading values. Interestingly, the statements in these questions were testing many of the concerns expressed with the "commercialization" of distance education. So, a negative correlation means that the students that were more satisfied with their online experience: (1) did not miss the interactions of a traditional classroom (Q43), (2) did not want more interactions (Q30), (3) did not think they would have learned more if they took the course in a traditional environment on-campus, and (4) did not exhibit the feeling of being lost in the class.

Technology and Value: The last factor derived seems to be the least homogenous, as it appears to have three disparate subsets: Value, learning and technology. Q46 asks the subject directly to rate the value of the course, while Q19 asks whether or not the subject would take another distance course. These two questions represent the value characteristic of the factor. Q38 and the derived variable, LOHITECH, show the technology characteristic of the factor. The way that the LOHITECH variable was derived provides a measure of the magnitude of technology used in the reference course, while the perceived quality of that use may be indicated in Q38. Finally, Q22 and Q37 ask for ratings on "learning" for groups and thinking "critically." Tying these together in a single construct, one can imagine that for distance courses, two components such as technology use and learning influence the notion of value. In support of the viability of this proposition, Table 3 shows the correlations of Q22, Q37, Q38 and LOHITECH to the fundamental value question Q46.

Twenty-two variables from the questionnaire proved significantly correlated to satisfaction. A factor analysis of those 22 variables extracted five possible constructs. These constructs were labeled: Interaction with the Professor; Fairness; Content of the Course; Classroom Interaction; and Value, Technology and Learning. Each of these constructs were based upon the key characteristics of the underlying questions. Table 4 shows the results of combining the ratings for the questions in each construct and correlating each of them to satisfaction. As can be seen from the table, the constructs hold up well as five indicators of satisfaction.

Impact on Distance Education Administration

As mentioned earlier, the organization, which is the school in this case, is a key stakeholder in the success of a distance education program. Distance education courses and programs are not only used for providing an alternative delivery method for students, but also to generate revenues for the offering unit/college/university. As the number of distance courses and programs increase at an

Table 3. Correlation of Learning and Technology Use to Value (Q46)

Area(Question)	Correlation	Significance
Learning(Q22)	.426	.000
Learning(Q37)	.211	.035
Technology use (Q38)	.296	.005
Technology use (LOHITECH)	.203	.042

Table 4.   Correlation of Final Constructs to Satisfaction

Construct	Correlation	Significance
Professor Interaction	.771	.000
Fairness	.695	.000
Course Content	.588	.000
Classroom Interaction	-.515	.000
Technology Use and Value	.624	.000

exponential rate, the necessity to enhance quality and revenues also takes prominence.

The administration of distance programs is especially challenging since, for example, in most cases you probably will rarely see your students. Any feedback from students is generally limited to questionnaires and complaints. The distance education administrators have the responsibility to ensure that both the faculty and students are satisfied with their respective teaching and learning experiences. This satisfaction translates into an overall requirement for higher quality distance courses and better retention of distance students. Table 5 suggests operational recommendations based upon this study that can impact online program success.

The data in this study indicates that a timely and personalized feedback by professors results in a higher level of satisfaction by students. The administrators, therefore, have to work closely with their faculty and offer them ways to enrich the teacher-student relationship. Paradoxically, the faculty need to use technologies to add a personal touch to the virtual classroom. For example, faculty should be encouraged to increase the usage of discussion forums, responding to email within 24 to 48 hours, and keeping students up-to-date with the latest happenings related to the course.

From an administrative perspective, fairness in the virtual classroom is an academic issue and, therefore, is best left to the instructor's prerogative. The pedagogical details such as assignments, grading scale, test and quiz content, etc., are best left to the instructor. However, administrators can work closely with faculty members to develop effective policies and procedures that help set realistic student and faculty expectations. For example, some content delivery platforms do not keep their session "active" while the student takes a test. In turn, some ISPs will log users off after 15 minutes of inactivity. This combination of automated tests longer than 15 minutes and an ISP inactivity policy that automatically logs users off after 15 minutes can create great havoc and anxiety for distance students. A simple administrator "heads up" to students and faculty about this possible technology quirk with the testing platform can help with the perception of fairness.

The data also indicates that good course content and explicit instructions increase student satisfaction in the virtual classroom. It may well be that course content basically sets expectations and explicit instructions manage these expectations for the distance student. This result suggests that faculty should have complete web sites with syllabi and detailed instructions. In turn, it also suggests that distance education administrators should focus their attention on providing faculty with support, such as good web site design, instructional designer support, test design, user interaction techniques, etc., appropriate for distance learning.

Interestingly, in a further analysis of this data, distance students that were more satisfied with their online experience didn't miss the traditional classroom interactions, did not think they would have learned more if they took the same course on campus, and did not feel that they were lost in the classroom. It is clear that a majority of students who enroll in distance classes are self-selected and know what to expect in an online course. They realize and accept that their experience will not be similar to a traditional classroom experience. In turn, they seem better prepared to make the most out of the distance classroom tools to enhance their learning experience. This data is very valuable to the distance administrator who is working on retention efforts. Knowing that these traits are significant for student satisfaction, an administrator can focus on and market to prospective students who are more likely to exhibit these traits.

Since distance students' notion of value intertwines learning and technology, it is imperative that distance administrators offer and that faculty use the available technology in the distance program. Technology, in this case, not only refers to the actual software and hardware features of the platform, but also how well technology is adapted to the best practices of teaching. The results imply that if technology is available, but not used, it lowers satisfaction. So, technology options that are not being used in a course should not appear available. For the

Table 5.   Recommendations to Increase Online Program Success

1.	Have instructors use a 24-48 hour turnaround for e-mail
2.	Have instructors use a one week turnaround for graded assignments
3.	Provide weekly "keeping in touch" communications
4.	Provide clear expectation of workload
5.	Provide explicit grading policies
6.	Explicitly separate technical and pedagogical issues
7.	Have policies in place that deal effectively with technical problems
8.	Provide detailed unambiguous instructions for coursework submission
9.	Provide faculty with instructional design support
10.	Do not force student interaction without good pedagogical rationale
11.	Do not force technological interaction without good pedagogical purpose
12.	Collect regular student and faculty feedback for continuous improvement

program administrator, this would suggest adoption of distance platforms that are customizable at the course level with respect to displaying technological options.

Conclusion

This study attempts to identify potential indicators for satisfaction with distance education. A body of potential indicators was derived from the literature surrounding the traditional-vs.-virtual classroom debate. A 49-question questionnaire was developed from the indicators and was administered to M.B.A. students in an established distance education program. A total of 101 usable questionnaires were analyzed, with the result that 22 variables correlated significantly to satisfaction. A factor analysis extracted five basic constructs and these factors were defined based upon the variables within each construct. Assuming that the constructs were valid, the chapter's final section discusses several administrative implications for obtaining more satisfaction in a distance program.

References

Aranda, E. K., Aranda, L. and Conlon, K. (1998). *Teams: Structure, Process, Culture and Politics.* New York: Prentice-Hall.

Bilimoria, D. and Wheeler, J. V. (1995). Learning-centered education: A guide to resources and implementation. *Journal of Management Education*, 19(3), 409-428.

Black, J. (1997). Retrieved January 12, 2000 from the World Wide Web: http://www.news.com/News/Item/0,4,7147,00.html.

Boyatzis, R. E. and Kram, K. E. (1999, Autumn). Reconstructing management education as lifelong learning. *Selections*, 16(1), 17-27.

Brookfield, S. D. (1995). *Becoming a Critically Reflective Teacher.* San Francisco, CA: Jossey-Bass.

Carr, S. (2000). As distance education comes of age the challenge is keeping students. *Chronicle of Higher Education*, February 11.

Chidambaram, L. (1996, June). Relational development in computer-supported groups. *MIS Quarterly*.

Creahan, T. A. and Hoge, B. (1998). *Distance Learning: Paradigm Shift of Pedagogical Drift?* Presentation at Fifth EDINEB Conference, September 1998, Cleveland, Ohio, USA.

Dede, C. (1996). The evolution of distance education: Emerging technologies and distributed learning. *American Journal of Distance Education*, 10(2), 4-36.

Dixon, D. (2001). *Mentoring Over Distance*. Unpublished doctoral dissertation, Fielding Graduate Institute.

Gobal, A., Bostrom, R. B. and Chin, W. (1993, Winter). Applying adaptive structuration theory to investigate the process of group support systems. *Journal of Management Information Systems*, 9(3), 45-69.

Gunawarden, C., Lowe, C. and Carabajal, K. (2000). *Evaluating Online Learning Models and Methods*. Society of Information Technology and Teacher Evaluation International Conference in San Diego, California. ERIC No.: ED444552.

Gustafsson, A., Ekdahl, F., Falk, K. and Johnson, M. (2000, January). Linking customer satisfaction to product design: A key to success for Volvo. *Quality Management Journal*, 7(1), 27-38.

Haythornthwaite, C., Kazmer, M. M., Robins, J. and Showmaker, S. (2000, September). Community development among distance learners. *Journal of Computer-Mediated Communication*, 6(1).

Hearn, G. and Scott, D. (1998, September). Students staying home. *Futures*, 30(7), 731-737.

Hogan, D. and Kwiatkowksi, R. (1998, November). Emotional aspects of large group teaching. *Human Relations*, 51(11), 1403-1417.

Hora, N. and Kling, R. (2001). Students' distress with a Web-based distance education course: An ethnographic study of participants' experiences. *Information, Communication & Society,* 3(4), 557-579.

Institute for Higher Education Policy. (2000). *Quality on the line: Benchmarks for Success in Internet Distance Education.* Washington, D.C.

Kerlinger, F. N. (1986). *Foundations of Behavioral Research.* 3rd ed. New York: Holt, Rinehart & Winston.

Kim, J. and Mueller, C. W. (1986). *Introduction to Factor Analysis*. Beverly Hills, CA: Sage Publications.

Kirkman, B. L., Rosen, B., Gibson, C. B., Etsluk, P. E. and McPherson, S. (2002, August). Five challenges to virtual team success: Lessons from Sabre, Inc. *The Academy of Management Executives*, 16(3).

Kline, P. (1993). *The Handbook of Psychological Testing*. London: Routledge.

Kling, R. (2000). Learning about information technologies and social change: The contribution of social informatics. *Information Society*, 16(3), 217-232.

Martz, B. and Shepherd, M. (2002, July). Using the influence level of information to explain the non-consensus process loss. *Group Decision and Negotiation*, 11(4).

McGrath, J. E. (1984). Dilemetics: The study of research choices and dilemmas. In J. E. McGrath, J. Martin and R. A. Kulka (Eds.), *Judgment Calls in Research*. Beverly Hills, CA: Sage Publications.

Moore, M. G. (1995). American Distance Education: A short literature review. In F. Lockwood (Ed.), *Open and Distance Learning Today*. New York: Routledge.

Noble, D. F. (1999). *Digital Diploma Mills*. Retrieved November 28, 2002 from the World Wide Web: http://www.firstmonday.dk/issues/issue3_1/noble/index.html.

Nunally, J. (1978). *Psychometric Theory*. New York: McGraw-Hill.

Nunamaker, Jr., J. F., Dennis, A. R., Valacich, J. S., Vogel, D. R. and George, J. F. (1991). Electronic meeting systems to support group work. *Communications of the ACM*, 34, 42-58.

Pallof, R. & Pratt, K. (1999). *Building Learning Communities in Cyberspace: Effective Strategies for the Online Classroom*. San Francisco, CA: Jossey-Bass.

Peters, T. J. and Waterman, Jr., R. H. (1982). *In Search of Excellence*. New York: Harper and Row.

Polanyi, M. (1966). *The Tacit Dimension*. Garden City, NY: Doubleday.

Poole, M. S. and DeSanctis, G. (1990). Understanding the Use of Group Decision Support Systems: The Theory of Adaptive Structuration. In J. Fulk and C. Steinfeld (Eds.), *Organizations and Communication Technology*. New Bury Park, CA: Sage Publications.

Ragan, L. C. (1998). Good Teaching is Good Teaching. Retrieved from the World Wide Web: www.ed.psu.edu/acsde.

Russell, T. (2001). Retrieved January 12, 2001 from the World Wide Web: http://nova.teleeducation.nb.ca/nosignificantdifference/.

Shepherd, M. M., Martz, B., Ferguson, J. and Klein, G. (2002, June). *A Survey of Distance Education Programs*. In Conference Proceedings of EDINEB IX, Guadalaraj, Mexico.

Statsoft. (2002). Retrieved November 30, 2002 from the World Wide Web: http://www.statsoftinc.com/textbook/stfacan.html.

Svetcov, D. (2000). The virtual classroom vs. the real one. *Forbes*, 50-52.

TeleEducation, NB. (2001). Retrieved January 12, 2001 from the World Wide Web: http://teleeducation.nb.ca/anygood/.

Tinto, V. (1975). *Leaving College.* Chicago, IL: University of Chicago Press.

Tu, C. (2000). Online learning migration: From social learning theory to social presence theory in the CMC environment. *Journal of Network and Computer Applications*, 23, 27-37.

Tucker, L. F. and MacCallum, R. (1997). Unpublished manuscript. Retrieved November 30, 2002 from the World Wide Web: available at http://quantrm2.psy.ohio-state.edu/maccallum/book/ch6.pdf.

U.S. News & World Report. (2001, October). Best Online Graduate Programs.

Vavra, T. G. (2002, May). ISO 9001:2000 and customer satisfaction. *Quality Progress*, 35(5), 69-75.

Appendix A - Distance Education Questionnaire*

You may receive multiple copies of this questionnaire. Please, complete only once.

Demographic: No course instructor will see the individual results associated with your name. Please answer the questions honestly so that the results can be relied upon as we make several important decisions regarding improving the UCCS distance program. Please identify one of the courses completed as your Reference Course. If this is your first course, list it as the Reference Course. When a question refers to a "course" please keep your Reference Course in mind.

Q1. Sex: [ ] M [ ] F

Q2. Reference Course: _____

Q2b. Student Type: _____

Q2c. Grade _____

Q3. How many distance courses have you completed (do not count those in which you are currently enrolled but have not completed)?

    ___ courses (0 if first)

Q4. The amount of time per week, I **expected** to spend on the course:

    [ ] 0-2 hrs [ ] 2-4 hrs [ ] 4-7 hrs [ ] 7-10 hrs [ ] 10+ hrs

Q5. The amount of time per week, I **did** spend on the course:

    [ ] 0-2 hrs [ ] 2-4 hrs [ ] 4-7 hrs [ ] 7-10 hrs [ ] 10+ hrs

Class Features: On a scale of Strongly Disagree (SD) to Strongly Agree (SA), please rate the following class features.

Class Features	Used (Y/N)	Rating (circle one)				
Q6 E-mail was effectively used	Y N	SD	D	N	A	SA
Q7 Forum (class wide) was effectively used	Y N	SD	D	N	A	SA
Q8 Forum (group wide) was effectively·used	Y N	SD	D	N	A	SA
Q9 Online timed tests were effectively used	Y N	SD	D	N	A	SA
Q10 Team/group projects were effectively used	Y N	SD	D	N	A	SA
Q11 Announcements were effectively used	Y N	SD	D	N	A	SA
Q12 Phone calls from professor were effectively used	Y N	SD	D	N	A	SA
Q13 Electronic gradebook was effectiyely used	Y N	SD	D	N	A	SA
Q14 Fax/postal service was effectively used	Y N	SD	D	N	A	SA
Q15 Other:	Y N	SD	D	N	A	SA

Survey: On a scale of Strongly Disagree (SD) to Strongly Agree (SA), please rate the following statements based upon your personal experience and opinion.

Statement	Rating (circle one)				
Q16 I was satisfied with the content of the course	SD	D	N	A	SA
Q17 The tests were fair assessments of my knowledge	SD	D	N	A	SA
Q18 I would take another distance course with this professor	SD	D	N	A	SA
Q19 I would take another distance course	SD	D	N	A	SA
Q20 The course workload was fair	SD	D	N	A	SA
Q21 The amount of interaction with the professor and other students was what I expected	SD	D	N	A	SA
Q22 The course used groups to help with learning	SD	D	N	A	SA
Q23 I would have liked to have had more interaction with the professor	SD	D	N	A	SA
Q24 Overall, I was satisfied with the course	SD	D	N	A	SA
Q25 I spent more time than expected on this course	SD	D	N	A	SA
Q26 The course content was valuable to me personally	SD	D	N	A	SA
Q27 The communication technology (Internet ISP, forum, modem, Jones system) did not disrupt the course	SD	D	N	A	SA
Q28 Grading was fair	SD	D	N	A	SA
Q29 Distance instructors are more lenient than on campus instructors	SD	D	N	A	SA
Q30 Often I felt "lost" in the distance class	SD	D	N	A	SA
Q31 The class instructions were explicit	SD	D	N	A	SA
Q32 I "stayed current" with the reading assignments	SD	D	N	A	SA
Q33 Feedback from the instructor was timely	SD	D	N	A	SA
Q34 I received personalized feedback from the instructor	SD	D	N	A	SA
Q35 I am very comfortable with using computers: Internet, upload/download of files, Word, Excel, and PowerPoint	SD	D	N	A	SA
Q36 I would have learned more if I had taken this class on-campus (as opposed to online)	SD	D	N	A	SA
Q37 This course made me think critically about the issues covered	SD	D	N	A	SA
Q38 I think technology (e-mail, Web, discussion forums) was utilized effectively in this class	SD	D	N	A	SA
Q39 I felt that I could customize my learning more in the distance format	SD	D	N	A	SA
Q40 I would call myself an introvert	SD	D	N	A	SA
Q41 AACSB accreditation was important to my choosing this program	SD	D	N	A	SA
Q42 The course content was valuable to me professionally	SD	D	N	A	SA
Q43 I missed the interaction of a "live," traditional classroom	SD	D	N	A	SA
Q44 I made friends in the course that I continue to communicate with	SD	D	N	A	SA
Q45 The reputation of UCCS was important to me in choosing this program	SD	D	N	A	SA
Q46 Overall, the program is a good value (quality/cost)	SD	D	N	A	SA
Q47 I participated more in this Reference Course than I have/had in traditional (on campus) courses	SD	D	N	A	SA
Q48 I chose the UCCS distance MBA because it does not have a campus attendance requirement	SD	D	N	A	SA
Q49 The technical support provided by Jones was good (leave unanswered if you never used technical support)	SD	D	N	A	SA
Comments:					

Note: The question numbers (along with variables "Q2b Student Type ___ Q2c Grade ____") are added for reference purposes.

Chapter VIII

Online Assessment in Higher Education: Strategies to Systematically Evaluate Student Learning

Elizabeth A. Buchanan
University of Wisconsin-Milwaukee, USA

Abstract

This chapter acknowledges the challenges surrounding assessment techniques in online education at the higher education level. It asks specifically, "How do we know our online students are learning?" To get closer to answering this question with confidence, various strategies ranging from participation techniques to online group work, peer and self-assessment, and journals and portfolios are described. The role of online mentoring as a supplementary strategy is also introduced. The chapter concludes with a survey of advantages and disadvantages of the various strategies.

Introduction: What Are Our Students Learning?

How do we know our students are learning? That is one of the most challenging questions facing educators in all levels of formal schooling, from elementary to higher education to training; assessing what learning is taking place proves to be a demanding task for all involved. On a continuum, we have such mechanisms as standardized tests on one extreme to looking at our students' faces on the other to ascertain whether "learning" is occurring. Over the years, various types of assessment models have been developed, each purporting to measure learning "better" and more authentically than its predecessor and, yet, whether we can still affirmatively answer than age-old question, "Are my students learning" remains elusive. Schunk is correct in noting, "assessing learning is difficult because we do not observe it directly but rather we observe its products or outcomes" (2000, p. 7). That is, learning itself is not quantifiable or tangible — we see test scores, read papers, and observe discussions, all of which supposedly gauge *something* happening between student and content. What that *something may be* is debatable. When we add to this ambiguity the complexities of online education, we have many issues and challenges to consider vis-à-vis student learning.

In higher education today, for a variety of reasons, distance education in the form of online or Web-based delivery has taken root as a popular, cost-effective, and pedagogically sound process of teaching and learning. Numerous studies over the years have sought to affirm that distance education is equally as effective (why not better?) as face-to face learning. Many studies have oftentimes concluded with the "no significant difference" phenomena, suggesting that there is no difference in the learning outcomes that occur in a distance environment versus on-site. That may be so, and our students may be "learning" all the same, but regardless of environment, a good educator will want to know *what* his or her students are learning and *how* do they process the information and learning objects from a course into *knowledge*. The good educator will want to know "what works" and why? Moreover, the good online educator will want to know how to best use the online environment to promote student learning and how to adapt old strategies and adopt new strategies to foster online learning. To get at the root of these questions, we must assess assessment strategies themselves in online environments. Are we looking at the right things to assess our students in online coursework? Are we as instructors asking the right questions? Are we engaging our students in the right activities?

This chapter will look at strategies to employ in assessing students in higher education level, Web-based coursework, though this author believes some of these apply to higher-level secondary schooling, as well. Some of these strategies work equally as well in on-site environments, too, and can be used with confidence in a variety of educational settings. This chapter does not address training in particular, however.

It is important to realize that when discussing assessment, discipline-specific perspectives are significant and can render any general discussion of assessment irrelevant. Readers are encouraged to pursue research in their disciplines regarding subject-specific assessment techniques. The goals of this chapter are to present a general discussion of assessment in online environments for instructors new to the online classroom, while moving educators closer to understanding and valuing student learning in online environments. The strategies described herein are intentionally devoid of a disciplinary context.

Assessing in Online Environments

We hear, all too often, about failures in education: failing test scores, illiteracy, social promotion, etc. And, we as educators may unfortunately have experienced failures in our classrooms, online or on-site, to some degree. What constitutes failure? Student grades? Student attrition? Student inability to understand or process information into knowledge? Students not knowing what *we think* they *are supposed* to know?

This last failure, in particular, is common in online learning for a variety of reasons: the disparate spaces of students and instructors often contributes to misunderstandings, technological failures or mishaps which contribute to student distress (Hara and ·Kling, 2001), poor instructional design so that students become lost or confused within a Web course environment (Berge et al., 2000) and, a mismatch between course and learning objectives to their assessment. Rogers (2000, p. 5) states: "… matching goals and objectives to assessment is good practice … the problem many teachers have is keeping the goals and the skills you taught to meet the goals, and how students' progress through the materials is related to assessment. Many times, assessment instruments do not measure what was actually taught."

Assessment can, and should, propel the design of an online course. By first asking two fairly simple questions, "What do I want students to learn in this course?" and "What types of learning activities will contribute to this?," online instructors are off to a good start. These two questions are interrelated with a larger issue: the instructor's pedagogical beliefs and perspectives. We have heard over and over in the distance education literature that the instructor's role changes from the sage on the stage to that of an active contributor, or guide, more in line with constructivist theories of education. Regardless of the formal theory one believes in, it is important for instructors to consider what they believe about teaching and learning. We can come to understand our pedagogies better and ultimately teach better by systematically investigating questions of teaching and learning. One way of promoting such systematic investigation is through an ongoing, constant evaluation of aims and objectives. This can be done at both the micro (lesson or weekly) and macro (course in general) levels. Online education, in particular, can facilitate this type of flexibility and openness to revision by its very nature.

When considering online assessment, one must reflect on instructional design, what to include in the course and how to present it. Berge, Collins and Dougherty (2000, p. 34) claim that, "instructional design focuses on analyzing what the course learning objectives are and how to best present them to students." Design principles must take into account the necessary elements requisite to an online environment: syllabus, schedule (or what Berge et al. call "administrivia"), lectures (in text, graphic, or video format), readings, discussion boards, synchronous chat function, quiz or test tool (if relevant), and additional resources or materials. These elements should be structured logically and easy for students to find and use. Rogers (2002) suggests designing not for the ideal, but for the reality of your situation, suggesting that to a great extent, online course design should be considered on a case-by-case basis. Additionally, "Material should be presented in such a way that it is compatible with a number of learning styles. *Visual* learners benefit most from charts, maps, filmstrips, notes and flashcards. *Auditory* learners benefit most from tapes, videos, lectures, notes, and recitation. *Tactile* learners benefit most from writing repetition, construction and display projects, note taking, analogy, and study sheets. Consider all three styles when designing course material and activities" (Madden, 1999).

The main goal of course design is to facilitate a learning environment where students can engage with the materials and course content in a meaningful way. Certain assessment strategies can promote this engagement extremely well.

Options in Online Assessment Strategies

By systematically examining issues of teaching and learning in online environments, instructors can develop meaningful assessment mechanisms that are fair and worthwhile. That is, they will tell us what we want to know about our students' learning. Assessment, in any environment, must be systematically applied and consistent. Instructors are encouraged to provide students with a detailed discussion of the assessment tools used *and* the philosophies behind those tools, so students understand why certain assessment mechanisms are in place. By understanding the types of assessments, students may see more clearly what is expected of them and how they are to achieve course and learning objectives.

Unfortunately, all too often, students receive little more than a grade or a number, and these marks do not encourage students to engage more deeply in their learning pursuits. Feedback from instructors in online environments is even more important than in onsite coursework, to avoid students' sense of isolation, confusion, and apathy toward faceless and far-removed instructors and peers. The following strategies encourage active relationships among students and instructors that provide different perspectives on learning. While tests and quantitative measures are, indeed, useful and have a place in online education, these strategies promote a more systematic investigation into learning that allow us to better answer that age-old question, "Are my students learning?" These strategies, too, however, require considerable commitment from instructors, but achieve the types of learning environment where one can know with greater certainty that students are learning.

Participation

Many Web courses rely on participation as a major component of assessment. A major problem arises when "participation" is not clearly defined or articulated. Instructors should provide guidelines as to their expectations of participation: Is it a quantitative measure? Qualitative? Both? Is participation simply fulfilling weekly assignments? Is it responding to other messages to continue a discussion theme? To simply state, "participate by commenting on the readings or discussion postings" is inadequate. If the instructor stops and considers, "What do I want to achieve by having students comment on the readings or participate on the discussion board?" or "What do I want students to get out of the readings?," participation will be redefined and refocused in a meaningful manner.

Typically, participation takes place in the form of discussion through asynchronous tools, such as bulletin boards. It is very important for instructors to consider the use of discussion, as Muilenburg and Berge (2002, p. 104) describe: "Does online discussion make sense? Is the discussion method a good match for the overall goals and objectives of the course, for the content, and for the learners? If the answer to any of these is 'no,' then you can spend all of your time ... with disappointing results. Use online discussion only when it makes a valuable contribution to the desired outcomes of the course ... ."

This instructor has found a useful approach to participation in the form of peer citations. Of particular interest, the asynchronous nature of online courses allows peer citation to be a valuable strategy in promoting meaningful participation.

Students can receive credit for citing others in their postings and assignments, as well as receive credit for being cited. Just as in peer-reviewed research where citations are valued, so should they be in online coursework. A reciprocal relationship grows out of the peer citation strategy, in that students strive to do their best work in order to be cited, and students respond enthusiastically to readings and discussions, looking for the best citations they can find.

Portfolios and Journals

Portfolios and journals are both reflective collections. Portfolios are collections of student work over time and can include such things as excerpts from a learning journal, papers, exams, in-class writing, online discussion postings, peer assessments, and the like, while journals, more specifically focus on students' impressions, feelings, thoughts, actions, surprises, frustrations, etc., to the class or learning situation. In the learning context, however, both reflect a learner's experience in a classroom. The North Central Regional Educational Laboratory (2002) says of portfolios, "Portfolios of student performance and products have gained support from educators, who view them as a way to collect evidence of a student's learning. For many educators, portfolios are an attractive alternative to more traditional assessment methods." In particular, journals and portfolios are extremely useful tools for gauging student progress, as they are longitudinal and, thus, reveal changes over time (usually throughout a semester). They do, however, require a substantial commitment on the part of both students and the instructor. Students must think carefully about their learning experience that day or that week, articulate their reactions, and spend time reviewing previous entries to see how they are progressing as learners. Also, on the instructor side, time spent reading and responding to student entries is significant. Portfolios or journals are only useful as learning assessments if the instructor commits to creating a dialogue with the students by asking questions of the entries, challenging assumptions, raising awareness, and so on, through the process. The learning journal is an interactive and "living" piece of assessment. Instructors, with extremely heavy student loads can, however, choose to sample student entries, which still allow an instructor to engage with the student sufficiently, but avoids the burnout of reading and responding to too many student entries.

Such tools have the potential to become invaluable if the students are not documenting their learning experience "correctly." As with other assessment techniques, journals in particular require explanation and guidance — students must be given guidelines to avoid "missing the point" in their entries. Brookfield (1995) has provided these prompts, which can be modified for use in an online environment:

- At what moment in class did you feel the most engaged with what was happening?

- At what moment in class did you feel most distanced or disconnected from what was happening?

- What action or incident that the instructor or other student took in class did you find the most helpful?

- What action or incident that the instructor or other student took in class did you find the most confusing?

- What happened in class this week that surprised you the most? (Brookfield, 1995)

This instructor has used these prompts in a hybrid course environment, and included such questions as:

- What role did the online portions of course play for you as a learner this week?

- Did the technology used impact you negatively or positively?

- Did the technology used this week inhibit or enhance your role as a learner?

- How did the technology impact your relationship with your peers this week?

By providing such guidelines, students have a starting point from which to evaluate their learning process. As noted, unique disciplines will want to ask particular questions to assess learning related to the subject matter in their own ways.

Self-Assessment

Related to portfolio or journal use is the technique of self-assessment, yet portfolios or journals are but two tools where students can self-assess. The use of self-assessment is a significant technique in online education and fits with the pedagogical specificity of Web-based learning in general by "shifting roles from audience to actors" (Cavanaugh, 2002, p. 183). Because students are "alone," self-assessment can promote more active engagement with the course than simply sitting back and awaiting a grade from one's instructor. Students must be comfortable with the idea of self-assessment, as it fails if students cannot be honest with themselves and the instructor. Schunk (2000, p. 379) advocates using self-assessment early in formal education, as early as preschool and kindergarten. He notes, "developing self-evaluation strategies helps students gain control over their learning. This, in turn, allows them to focus more effort in studying those areas where they need more time."

In short, with self-assessment, we are asking students to determine where they are as learners vis-à-vis some content. This is not always an easy task, for students may not know what they don't know, of course. To help alleviate this problem, the instructor must provide checklists, rubrics, or inventories to help students assess themselves. For disciplines more used to quantitative measures, pretests or practice tests can serve as means of self-assessment where discrete answers must be achieved.

Self-assessment must be guided by clear objectives. Students cannot adequately assess themselves if they are unclear about the goals and objectives of the lesson or content. Here, again, it is imperative for the instructor to articulate what he or she wants students to gain from a particular exercise or reading. Students' self-assessment also creates a dialogue with instructors whereby they can modify or revise their teaching. Self-assessment, as a formal assessment technique, opens the door to a more fluid teaching and learning environment, which coincides nicely with the structure of online environments.

Peer-Assessment and Group Work

The use of peer assessment usually raises fear among students, who are afraid of facing their friends the next week in class after issuing someone a bad grade. In online environments, part of this fear is minimized by the disparate nature of the students in Web-based courses, raising the utility of this assessment technique. Yet, because of this dynamic, another potential problem with peer assessment is the likelihood of inflated grades — no one wants to be known as the guy who gives harsh grades to his online friends, after all, so a higher-than-deserved grade may be issued. Whether individuals are harsher in virtual environments is debatable. Peer assessment does, however, have the ability to encourage students to strive harder to complete assignments and participate more actively if they know their peers are grading them. Thus, by its very nature, peer assessment may raise the bar of student engagement.

Peer assessment online can take shape in a number of ways. Students can assess each other's participation in online discussions, both synchronous and asynchronous, on assignments, and on group projects. (Group work will be discussed momentarily.)

Burgess and The Learning and Teaching Support Network (n.d.) suggests using the following points in peer assessment:

• It is best for students to make comments about their own strengths and weaknesses before hearing/receiving other people's views.

- Identify strengths of work first, before moving on to areas for improvement.

- Try to make comments descriptive rather than evaluative.

- Give feedback based on concrete behavior/examples rather than giving a general impression.

- Back up comments with evidence.

- Make sure the feedback can be clearly understood.

Instructors must be actively involved with peer assessment as well. They do not merely drop out of the learning experience and allow their students to complete the grading tasks. The instructor's role of the mediator takes effect with peer assessment, as he or she must first provide guidance to the students as to assessment criteria and then work with both the assessor and the assessed to be sure the assessment was fair and systematic. Peer assessment takes time to work extremely well in online classes and novice instructors should use this with some caution. Students must receive "training" on peer assessment. For valuable guidance, see Boud, Cohen and Sampson (2001).

Peer assessment becomes extremely valuable in group work. Typically, two major problems surround the use of group work: (1) given their many responsibilities of job, family, and school, it is difficult for students to find time to meet in groups outside of class, and (2) students are leery of having their individual grades dependent upon the abilities and efforts of others.

While online students will typically not meet synchronously and physically due to their disparate locales, they will be meeting electronically, through email or discussion forums. Students must worry, in this scenario, of nonresponsive peers. This instructor has heard student complaints about the frustrations of trying to contact other students via email or within course modules, only to be ignored. The second concern is quite valid for both online and onsite environments. Students complain that someone invariably fails to contribute equally or fairly. To prevent this, when using group work online, students should use peer assessment. Then, when a group member is unresponsive, the rest of the group need not worry that their grade will reflect this person's apathy because they, in fact, can control their own fate through peer assessment. Because peer assessment requires guidance from instructors, evaluation forms or checklists should be provided to students.

A simple evaluation form could include:

1. _____ participated actively in our group planning.

    Yes ☐ No ☐

2. _____ was willing to offer suggestions for our group work.

    Yes ☐ No ☐

3.   _____ was prepared for our group's working session.
       Yes ☐ No ☐

4.   _____ did his/her share of the work (typing, taking notes, etc.)

5.   I feel _____ contributed poorly/ adequately/ superiorly to my group.

6.   On a scale of 1-5, with 1 being the lowest and 5 the highest, _____ deserves
       a _____ for his/her work in the group.

7.   Think of a pie. You must slice up the pie based on the contributions of each
       member, with the biggest contributor getting the most pie. List each group
       member with a percentage of the pie. Be sure to include yourself in the
       slices.

These aforementioned strategies are a starting point for online instructors
interested in gauging student learning in different ways. They do, as noted,
require substantial effort on the part of both the instructor and the students, but,
in the long run, provide excellent data to document student learning.

Online Mentoring

Online mentoring is not a direct assessment strategy, but another means through
which the student experience, student progress, and learning itself, can be
gauged. Buchanan (2002) has suggested that online programs include a peer
mentoring network. The idea is simple: Students entering online programs are
matched with an advanced distance education student. Registration personnel or
advising staff within a department or school can facilitate the exchange of names
and e-mail addresses. The mentor and the mentee can communicate at their own
convenience and on their own terms. How does this promote learning or the
assessment of learning? In the general sense, institutions can gain knowledge
about the student experience by offering the mentor and the mentee short
surveys inquiring about the mentor relationship and what it contributed to each
individual. How does it help a novel online student assimilate to the online
environment? Will such a relationship help minimize student attrition? In the more
specific sense of student learning, instructors can inquire of the mentor his or her
perspective on his or her mentee's progress in courses. The mentor will have the
responsibility to ask basic questions of the mentee regarding his or her work in
specific courses: What learning strategies are being used in these courses? How
are mentees responding to these strategies? Basically, the mentor is getting a
clean look at how individuals respond to various assessment tools and sharing this

information with those responsible for promoting learning — the instructors themselves.

Online peer mentoring is a valuable and straightforward process that can contribute greatly to what Maki (2002) calls institutional curiosity:

Institutional curiosity seeks answers to questions about which students learn, what they learn, how well they learn, when they learn, and explores how pedagogies and educational experiences develop and foster student learning. When institutional curiosity drives assessment, faculty and professional staff across an institution raise these kinds of questions and jointly seek answers to them, based on the understanding that students' learning and development occur over time both inside and outside of the classroom.

Summary of Strategies and Advantages/ Disadvantages

This figure summarizes some of the existent advantages and disadvantages to the strategies discussed. It is noteworthy that the same point can be considered both simultaneously a pro and a con, depending on one's environment and nature of the course in particular. It is imperative that the instructor of an online course monitors and engages in the assessment techniques very closely to avoid the emergence of some of these disadvantages.

Conclusion

Ultimately, as Angelo (1999, n.p.) asserts, "Assessment efforts have resulted in little learning improvement because they have been implemented without a clear vision of what 'higher' or 'deeper' learning is and without an understanding or how assessment can promote such a vision ... we need a vision worth working toward ... we need a different concept of assessment itself, a new mental model ... and we need research-based guidelines for effective assessment practice that will increase the odds of achieving more productive instruction and more effective learning." Online education provides us with many opportunities to implement novel and innovative assessment strategies. As with any good online

Figure 1.

Strategy	Advantages	Disadvantages
Online participation	• Can be quantified • Can be analyzed for intellectual synthesis • Can encourage higher quality work and deeper engagement with course materials	• Can be quantified • Can be time intensive for instructor and students if not appropriately defined and explained
Portfolios and journals	• Show progression over time (longitudinal) • Large amounts of data • Promotes engagement and reflection • Students can demonstrate learning in a number of ways through portfolio/journal content • Can facilitate greater dialogue between student and instructor	• Typical university semester structure may be too short to see change longitudinally) • Instructors not knowing what parts reveal "learning" • Time intensive with large classes • Impetus on instructor to provide clear and articulate guidelines • Possibility of students missing the point of the exercise
Self-assessment	• Greater autonomy and sense of control exerted by student • Ability to see where one needs to improve • Facilitates dialogue between student and instructor on student level • Instructor must be clear in articulating goals and objectives to students	• Students may be more concerned with letter grade than learning process • Students may not know what they don't know • Instructor must provide guidelines/rubric • Instructor must be clear in articulating goals and objectives to students
Peer-assessment and group work	• May encourage students to work harder to "face" their peers • Can create a more collegial environment • Enables instructor to facilitate and oversee learning process among the students	• Possibility of grade inflation OR • Possibility of harsher grading • Can create a more hostile environment • Possibility of unfair and inequitable distribution of work and tasks if peer grading on group work is not used

course, the instructor must prepare ahead of time, be ready to revise and respond to students, and be ready to make a sound commitment to the learning experience. As educators, we must rise to the challenge to provide true learning opportunities for all.

References

Angelo, T. (1999). *Doing assessment as if learning matters most.* Retrieved November 27, 2002 from the World Wide Web: http://www.aahe.org/bulletin/angelomay99.htm.

Berge, Z., Collins, M. and Dougherty, K. (2000). Design guidelines for web-based courses. In B. Abbey (Ed.), *Instructional and Cognitive Impacts of Web-Based Education,* pp. 32-40. Hershey, PA: Idea Group Publishing.

Boud, D., Cohen, R. and Sampson, J. (2001). *Peer Learning in Higher Education: Learning From & With Each Other.* London: Kogan Page.

Brookfield, S. (1995). *Becoming a Critically Reflective Teacher.* San Francisco, CA: Jossey Bass.

Buchanan, E. (2002). (2002). Institutional and library services for distance education courses and programs. In R. Discenza, C. Howard and K. D. Schenk (Eds.), *The Design and Management of Effective Distance Learning Programs,* pp. 141-154. Hershey, PA: Idea Group Publishing.

Burgess, H. and Social Policy and Social Work/Learning and Teaching Support Network (LTSN). (n.d.). *Self and peer assessment.* Retrieved January 5, 2003 from the World Wide Web: http://www.swap.ac.uk/learning/Assessment2.asp.

Cavanaugh, C. (2002). Distance education quality: Success factors for resources, practices, and results. In R. Discenza, C. Howard and K. D. Schenk (Eds.), *The Design and Management of Effective Distance Learning Programs,* pp. 171-189. Hershey, PA: Idea Group Publishing.

Hara, N. and Kling, R. (2001). Student distress in web-based distance education. *EDUCAUSE Quarterly Articles,* (3), 68-69.

Madden, D. (1999). *17 elements of good online courses.* Retrieved January 5, 2003 from the World Wide Web: http://www.hcc.hawaii.edu/intranet/committees/FacDevCom/guidebk/online/web-elem.htm.

Maki, P. (2002). Developing an assessment plan to learn about student learning. *Journal of Academic Librarianship, 28*(1/2), 8-14.

Muilenburg, L. and Berge, Z. (2002). Designing discussion for the online classroom. In P. Rogers (Ed.), *Designing Instruction for Technology-Enhanced Learning*, pp. 100-113. Hershey, PA: Idea Group Publishing.

North Central Regional Educational Laboratory. (2002). *Self-assessment in portfolios*. Retrieved January 5, 2003 from the World Wide Web: http://www.ncrel.org/sdrs/areas/issues/students/learning/lr2port.htm.

Rogers, P. (2002). Teacher-Designers: How teachers use instructional design in real classrooms. In P. Rogers (Ed.), *Designing Instruction for Technology-Enhanced Learning*, pp. 1-18. Hershey, PA: Idea Group Publishing.

Schunk, D. (2000). *Learning Theories: An Educational Perspective*. Upper Saddle River, NJ: Prentice Hall.

Chapter IX

Assessing the Impact of Internet Testing: Lower Perceived Performance

Wm. Benjamin Martz, Jr.
University of Colorado at Colorado Springs, USA

Morgan M. Shepherd
University of Colorado at Colorado Springs, USA

Abstract

This chapter provides the results of a comparison between two sections of a graduate programming class, where one was an on-campus class and the other, a distance class. The course content, instructor, syllabus, lecture materials, notes, assessments and semester (time of year) were the same. Both groups were surveyed to test their satisfaction with the testing procedure and with their perception of certain aspects of the social environment. The results showed differences in perceived test performance. Two conjectures about possible causes underlying the difference and suggestions for possible future research end the discussion.

Introduction

Distance education is big business; over 1.6 million students enroll in distance education classes annually. The Institute of Higher Education (2000) predicts that by 2004 over 90% of all two and four-year colleges will offer some sort of online courses. By the end of 2003, the virtual education market will grow in excess of $21 billion (Svetcov, 2000). New technologies and improvements in networking capabilities are enabling distance education instructors to come closer to providing the traditional learning environment for their students. However, there are still many issues that need to be resolved and, according to recent research studies, it does not appear that these will be resolved anytime soon.

Interestingly, the research in distance education goes back to the late 20's when the first studies were published comparing the test scores of students in a classroom to their counterparts in a correspondence course (Crump, 1928). Since then, hundreds of journal articles, studies and reports have been published with similar comparisons with TV, radio, video tapes, computer-based training, audio-conferencing, groupware, and now the Internet, representing the technology compared to the traditional classroom (Moore, 1995).

Much of the research compares the distance education approach to the traditional classroom approach, looking for areas where the results from the distance education approach equals or exceeds those from the traditional classroom approach. There are many confounds involved, making this type of research difficult. One of these confounds is the definition given to "distance education." These definitions range from correspondence courses, to satellite classrooms where the instructor travels to lecture to a group of students meeting face-to-face, to courses that are held via e-mail, to courses that are held via two-way full motion video with other technological support. The definition debate will probably never end, as some definitions do not include the use of any technology (correspondence courses), while others require several technologies to be implemented (full motion video with chat rooms, list-servers and e-mail). For the purpose of this study, our definition of distance education involves a student body who never see each other or the instructor, who communicate via e-mail, phone or chat, and who hand in assignments via e-mail or via posting to a common work group area.

Other confounds arise due to the nature of the dependent variables that are studied and the interaction affects between them. Some of the dependent variables studied have been student performance, student satisfaction, and student retention. Within each of these three variables, socialization is thought to play a significant role. For example, Kling (2000) defines the study of complex issues around the interaction of information technologies with an institutional and cultural perspective as "social informatics." With the proliferation of the Internet, distance education is becoming a legitimate arena for social informatics. Kling's definition guides us to look at social issues brought on by the use of the Internet technology with education in general.

Other researchers (Shipley and Veroff, 1952; Papert, 1980; Hills and Francis, 1999) suggest that learning requires a social context to be effective. Bandura (1977) relies heavily on the interaction of people with their environment in his Social Learning Theory. Gunawardena and Little (1997) use results from their study to conjecture that "social presence is a strong predictor of satisfaction" in a computer-based environment. Tu (2000) goes even further by suggesting that the main driver of learning is the "consciousness of another person in the environment." Each of these studies indicates the importance of social issues in distance education. However, for each study that claims to show a significant difference between the distance and the traditional classroom environment, there seems to be one that claims to show that there is NO significant difference between the two environments. TeleEducation NB has helped compiled two lists of research with regard to distance education. Russell (2003) summarizes those studies that found "no significant difference" between students in distance education and students in traditional classroom environments. Those findings demonstrating a significant difference are available through the TeleEducation (2003) web site.

 Some studies have found significant differences in favor of the distance learning environment. For example, in a study on participation, Colorito (2001) found online students participating more than traditional students. Neslar and Hanner (2001) surveyed students from different nursing programs about their level of socialization and were surprised to find online nursing students showed more socialization characteristics than their peers in the traditional classroom environment. Gagne and Shepherd (2001) found online students less satisfied than their peers in the traditional classroom with the availability of the instructor in a distance environment. However, this does reinforce the importance of some type of social interaction to distance education students. Maki et al. (2000) found satisfaction with a lecture course to be lower for online students than for traditional students. Chen et al. (1991) found that the attitudes of students in a computer-based class were less positive than those attitudes from students in a conventional classroom.

By design, much of distance education is a solitary experience. In this light, Hogan and Kwiatkowski (1998) argue that the emotional aspects of teaching large groups with technology have been ignored. Hearn and Scott (1998) concur and suggest that before adopting technology for distance teaching, that technology must be able to address the social context of learning. This idea is somewhat borne out by Hill and Francis' (1999) work with Computer Based Training (CBT) wherein they found that students were more successful in the CBT environment when more social interaction occurred.

Several researchers suggest that there may be problems inherent with the distance environment. The most bothersome may be student retention where course drop out rates as high as 50% are reported (Carr, 2000). Brown and Liedholm (2002) found that students in the virtual classroom performed worse than their traditional counterparts on examinations. Without the hands-on supervision inherent in the traditional environment, cheating and plagiarism take on added concerns (Agger-Gupta, 2002).

Other social factors also are being examined. Kirkman et al. (2002) identified two social factors, creating trust and the feeling of isolation, as key factors for students to succeed in the distance environment. Hora and Kling (2002) research and report on the concern that distance courses may create new anxiety and stress for the student. Haythornthwaite et al. (2000) found that the distance environment limited the number of social cues and, in turn, this limitation reduced the participation found in a distance environment.

This study looks more closely at student satisfaction, specifically student satisfaction with the testing process in our distance education program. To help reduce the threat of cheating, all student tests are timed. A timer starts when the student begins the test. The timer counts down the time and is clearly visible in the corner of the screen. After the timer is started, the student can view the questions and take the test. Should the timer expire, the student's work up to that point is automatically submitted.

One potential problem with this method is the increased stress that might arise. Building on the work of Hora and Kling (2002) and Haythornthwaite et al. (2000), we hypothesize that this stress comes from two sources. The first source is the fear of not performing well. The second source is the fear of not performing as well as your peers. Both fears exist in a traditional classroom, although the latter fear may be somewhat lessened due to the fact that the student can receive some visual clues as to how well they are doing compared to their peers.

In the traditional test-taking environment students get feedback from the other students in the room. This feedback takes several forms. The first is in the form of a pre-quiz, "working" of the room. Students ask each other how much they studied for this quiz and how difficult or easy they think it will be. A second form of feedback occurs during the actual taking of the test. In the role of a student, suppose I see that time is running out and I'm only 75% of the way through the quiz. If I notice that 90% of my peers are still taking the quiz I may realize that at least we are all "tanking" this quiz together (misery loves company). This socialization cue will help to dampen my fear. Yet another cue is in the form of the post-quiz "working" of the room. Students get a chance to see how well the others thought they did (albeit perceived performance) on the quiz. The distance student does not have easy access to any of these socialization cues. Hence, we decided to explore the effects of these socialization aspects on student satisfaction in the distance test-taking environment.

We hypothesized that due to this lack of socialization cues in the distance education environment, the distance students would be less satisfied with their performance on the quiz than the traditional classroom students. We devised the following experiment to test this hypothesis.

The Research Study

Like many of the other studies discussed earlier, this research study was undertaken to explore the possibility of differences between the traditional classroom environment and the distance education environment (Figure 1). However, the unit of analysis for this study is the common learning activity of test taking. Most studies noted earlier use the complete course as the unit of study, i.e., student performance over the span of an entire course. This study uses a subset activity of a course, a test. The rationale of using this lower level of granularity is that differences that may occur at the test level may well be "washed out" when the aggregate unit of measure, the course, is used.

The subjects in this study were drawn from two Masters of Business Administration (MBA) level classes. Both classes were introductory information systems classes teaching Visual Basic© programming. One was offered through an established distance program while the second was taught in a traditional classroom. Both courses were offered in the same 16-week semester format, by

Figure 1.   Research Design

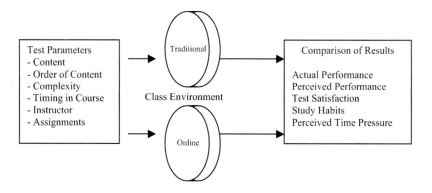

the same instructor, using the same book, same lecture material including slides, same assignments and the same grading scale. The weekly assignments and order of topics was the same. The distance class had nine students and the on-campus class had 10 students.

The actual test used in the study was the first test of the semester for both classes. The distance class took the test using an automated, online testing program provided by the course content package that was accessed over the internet. The on-campus class took its test on paper. The tests had the same questions with the same weights and the same order. The test was divided into three sections so as to minimize the potential problems in the distance environment. Both environments imposed the same time limit. The online test automatically enforced the time limit, while a timer was used to enforce the on-campus class time limit.

A 15-question questionnaire was developed using the themes of interest from previous distance education research. Questions were developed to explore the areas assumed to be fundamentally different between distance and traditional classroom education. The questions were developed to look for potential differences in actual performance, perceived performance, test satisfaction, study habits, and time pressure.

In both cases, immediately after the test, the questionnaire was distributed to the class members. The distance class received the questionnaire as an e-mail attachment. All questionnaires were returned within three days. The on-campus class filled out and turned in the questionnaire before leaving the classroom. All students in both classes returned their questionnaires.

Results

The responses from the two groups were tabulated and are shown in Table 1. In addition, Table 1 shows the significance of a statistical test, the Fisher Exact Test (Siegel, 1958), comparing the results of the online students to the on-campus students for each question.

The first important observation is that two of the questions (Q4 and Q12) show a significant difference. Q4 hints at a social interaction difference between the two groups. Clearly, online students do not want to take tests with other students around, while the on-campus students seem to prefer to have other students around.

Question 12 proves a little more interesting. Q12 shows that the two groups of M.B.A. students report a significant difference in their satisfaction with their

Table 1.   Comparison of Online to On-Campus Groups

Statement	Online		On-Campus		Sign. (1)
	A	D	A	D	
Q1. I felt a lot of time pressure in taking this test.	3	6	2	8	.444
Q2. The time pressure helped me to perform well on this test.	0	9	2	8	.263
Q3. I would have scored higher with more time for the test.	4	5	1	9	.173
Q4. I prefer taking tests with my other classmates in a traditional classroom.	**0**	**9**	**6**	**4**	**.008**
Q5. I prefer taking tests by myself without other classmates in the room.	6	3	3	7	.128
Q6. The test was a fair way to evaluate how well I knew the subject matter.	6	3	9	1	.360
Q7. I studied for this test -- more than I usually do for tests.	0	9	1	9	.526
Q8. I studied with others (study group, non-classmate, etc.) for this test.	1	8	0	10	.474
Q9. I prefer studying with others for tests.	3	6	2	8	.444
Q10. I believe that I finished sooner than 50% of the others taking this test.	2	7	5	5	.220
Q11. I believe that I scored better than 50% of the others taking this test.	2	7	5	5	.220
Q12. Overall, I was satisfied with my performance on the test.	**2**	**7**	**8**	**2**	**.019**
Q13. Overall, I was satisfied with the format of the test.	6	3	8	2	.444
Q14. I waited until the last possible time to turn my test in.	3	6	1	9	.249
Q15. I studied between the sections of the test.	2	7	0	10	.211
Non-Questionnaire Objective Measures					
Correct Prediction of Position (CPP)	5		8		.259
Correct Prediction of Score (CPS)	5		4		.414
Actual Test Score -- ATS (average)	127		133		.211(2)
(1) Fisher's Exact Test (2) Mann-Whitney Test - (Siegel, 1958)					

own performance. Remember, this is really perceived performance, as the results were not yet returned in class or available online. This finding becomes more noteworthy because the actual test scores were not significantly different.

Questions 1, 2 and 3 deal with the issue of time pressure. Did the two groups feel pressure to complete the test? No, not really. Both groups disagreed that they felt time pressure and they disagreed that the time pressure helped their performance. While not statistically significant, the results for Q3 show that more of the online students would have preferred more time. Combined, these results show that a time limit did not induce the feelings of pressure, but the online group would better receive an extension of the time limit.

Questions 4 and 5 tried to determine differences in the social aspect of having others in the room when taking the test. Certainly, being able to see when others complete and hand in the test could induce comparisons. In turn, these comparisons, if negative, could induce higher levels of anxiety. The results of these questions track well with what would be expected. Those students that have chosen a distance education class report little interest in having other students as a reference group and those choosing an on-campus environment like having other students as reference.

Question 6 asks directly for an assessment of the fairness of the test. Both groups thought the test was a fair evaluation. This can be interpreted to mean that neither the technology nor the test seemed to impact on the students' perception of fairness.

Questions 7, 8 and 9 concentrate on the students' study habits. Questions 8 and 9 try to determine if there is a difference in a student's preference for studying with others. There was none. Question 7 was asked to see if there was any difference in perceived difficulty of the test. Neither group studied more than they normally do. This is significant for the online group, as it implies that they did not envision the distance exam to be harder than what they were used to.

Questions 10 and 11 were designed to test for the potential social impacts of the on-campus environment. As suspected, the on-campus students did better than the distance students when they were asked to predict in which half, top half or bottom half, of the class (1) they had completed and handed in their test, and (2) they had scored. The on-campus students were very accurate with their assessment of whether they turned their test in with the upper 50th percentile, or the lower 50th percentile. Half estimated that they finished in the first 50th percentile, and half estimated that they finished in the second 50th percentile. This indicates that the social cues were providing significant feedback. It is also interesting to note that their estimate as to whether they scored in the upper 50th percentile or the lower 50th percentile was 50-50.

Questions 12 and 13 deal with a student's satisfaction. The first question looks at his or her performance and the second asks about the format of the test. As stated earlier, the perceived performance proved significantly different between those students in the on-campus and those in the online class. This difference does not carry through on Q13, as no difference was found related to the format of the test.

Questions 14 and 15 compare each student's behavior with respect to when they handed in the test and to whether or not they studied between the sections of the test. One could expect the on-campus students to wait until the last moment, while the distance students do not. The results imply otherwise. Question 15 is truly unique to the distance environment. It indicates that two students found it necessary to study between the sections of the test.

Discussion

The first major limitation of this study is the sample size. With only 19 subjects spread over two classes, any differences found between the two groups will be sensitive to statistical prejudices. It should be noted that the Fisher Exact Test used to designate any differences as significant is the appropriate test for analyzing a two-by-two result. Also, it is a very conservative test, meaning that if a significant difference is noted, it is more likely a true difference.

In the counter-argument, one that argues for the reliability of the results, many of the typical confounds are not found in this study. For example, all subjects in both classes returned the survey. The results represent the whole population and not just a sample. Additionally, the courses were the same, the instructor was the same, the MS PowerPoint© and lecture notes were the same, the grading was the same and the test questions were in the same format and in the same order.

Notwithstanding the issues of a small sample size, the results still provide some interesting discussion points. First, why would there be a difference in perceived performance? A further analysis of the data did not find any of the other questions (Q1 through Q15) significantly correlated to Q12. So, none of the typical concerns looked at, such as "time pressure (Q1-Q3)," "fairness of the test (Q6)," "study habits (Q7-Q9)," or "format of the test (Q13)," help us understand the perceived performance difference exhibited.

However, one interpretation of Q10 and Q11 may provide a hint. While the differences between the subjects in these two classes are not statistically significant, it is interesting to note that the estimates of finishing position in the on-campus class better match the actual results than do the estimates in the distance class. This hints that the students in the on-campus class have a better reference point. Obviously, they can see the other students handing in their exams and compare their relative position. This reference point was not available to the distance class. Could it be that the reference point provided in the traditional classroom helps students understand their relative position and lead to better "perceived satisfaction"? This might also explain the decidedly negative impression that the distance students felt, concerning both whether they turned in their test in the upper or lower 50th percentile and whether they scored in the upper or lower 50th percentile.

The second area of interest is with time pressure. Again, without statistical significance, this is conjecture only. The numbers for the time pressure questions hint at time pressure impacting the online students more than the on-campus students. This is especially true when a student was asked if they would have scored higher with more time (Q3). The two groups had the same amount of time. But, we need to realize that a key difference was the "enforcer" of the deadline. In the distance case, the software counts down using a displayed clock and when zero is reached, the software closes down the test and locks the student out. This is a very rigid and impersonal activity in that there is no arguing with the computer. In the classroom environment, the "enforcer" is a human being. A student may feel it easier to push the time limit. Even as the instructor is saying, "It is time to turn the tests in," a student may feel comfortable finishing a sentence. Extrapolating then, could it be that the "impersonalness" of the automated testing environment influences perceived pressure?

The results raise one concern and provide two conjectures for understanding differences between on-campus and distance environments. The concern is about why online students would report lower perceived performance than on-campus students. Several keys factors suggested by the literature, such as test fairness and study habits, did not help to explain the difference. However, one conjecture consistent with the data is that an on-campus student has a built-in reference group when handing in tests: those students that have already handed the test in before them. The second conjecture consistent with the data is that "time pressure" may be inherent to the distance testing environment simply because the testing software is an impersonal enforcer of the time limit.

Lack of social cues may very well account for the more pessimistic attitudes of the distance students, even though there was no significant difference on the test. Now we need to look at various ways in which we can infuse the distance education environment with similar social cues. One way may be to put an "average time to complete quiz" number up on the screen for the distance students when they start their quiz, along with a "you are the XXth person to take this test" number so the students can better make sense out of the two. We may need to hold a special chat session prior to the quiz to allow the distance students to get some of these socialization cues. If we can infuse some portion of these social cues into the distance environment, we may be able to improve the satisfaction levels of the distance education students taking timed quizzes.

Conclusion

The distance education environment is big business. As such, there is much interest in understanding any differences inherent between these environments and traditional on-campus learning environments. The literature has a long list of researchers concerned with understanding these differences. Many hard, objective issues have been raised, such as performance, retention and participation. Lately, softer, subjective social issues, such as time pressure, anxiety, stress, and trust have been thrown into the debate. All in all, the results have been ambiguous. There are studies showing significant differences between the two environments and there are studies showing no significant differences. The latter imply that distance environment is "just as good" as the traditional environment.

The study reported here adds one more issue to the significant difference side of the equation. Using two classes, equivalent on many of the confounding variables, such as instructor, test format, time constraints, etc., data was gathered on issues such as perceived performance, study habits, and actual performance. The limitations of a small sample size notwithstanding, the results show that the distance students rated their perceived performance significantly lower than the on-campus students.

The attempt to interpret the implications of these findings suggested two conjectures consistent with the data. One, the testing process in the distance environment uses more of an impersonal enforcer of the time limit. As such, the distance student may feel more time is needed and feel less positive about his or her performance as the time limit runs out. Two, the on-campus environment provides students with a built-in reference group for analyzing their performance. As students hand in tests, they "know" where they fall in relation to others that have and have not yet handed in the test. Possible methods for adding the reference group to distance classes and for removing the personal confound from the on-campus environment were suggested.

The results of this study suggest one more area of concern for those looking to implement a distance environment: students in distance environments may exhibit lower perceived performance.

References

Agger-Gupta, D. (2002). Uncertain Frontiers: Exploring ethical dimensions of online learning. In K. E. Rudestam and J. Schoenholtz-Read (Eds.), *Handbook of Online Learning*. Sage Publishing.

Allen, M., Bourhis, J., Burrell, N. and Mabry, E. (2002). Comparing student satisfaction with distance education to traditional classrooms in higher education: A meta analysis. *American Journal of Distance Education*, 16(2).

Bandura, A. (1977). *Social Learning Theory*. Englewood Cliffs, NJ: Prentice-Hall.

Black, J. (1997). Retrieved March 30, 2003 from the World Wide Web: http://www.news.com/News/Item/0,4,7147,00.html.

Brown, B. W. and Liedholm, C. E. (2002, May). Can Web course replace the classroom in principles of microeconomics? *American Economics Review*.

Carr, S. (2000, February). As distance education comes of age the challenge is keeping students, *Chronicle of Higher Education*.

Chen, H., Lehman, J. and Armstrong, P. (1991). Comparison of performance and attitude in traditional and computer conferencing classes. *The American Journal of Distance Education*, 5(3), 51-64.

Colorito, R. (2001, September/October). Learning online. *Indiana Alumni Magazine*, 64(1).

Crump, R. E. (1928). *Correspondence and Class Extension Work in Oklahoma*. Doctoral Dissertation. Teachers College, Columbia University.

Gagne, M. and Shepherd, M. (2001, April). Distance learning in accounting: A comparison between distance and traditional accounting class. *T.H.E. Journal*, 28(9).

Gunawardena, C. H. and Zittle, F. J. (1997). Social presence as a predictor of satisfaction within a computer-mediated conferencing environment. *American Journal of Distance Education*, 11(3).

Hara, N. and Kling, R. (2001). Students' distress with a Web-based distance education course. *Information, Communication and Society*, 3(4).

Haythornthwaite, C., Kazmer, M. M., Robins, J. and Showmaker, S. (2000, September). Community development among distance learners. *Journal of Computer-Mediated Communication*, 6(1).

Hearn, G. and Scott, D. (1998, September). Students staying home. *Futures*, 30(7), 731-737.

Hills, H. and Peter, F. (1999). Interaction learning. *People Management*, 5(14), 48-49.

Hogan, D. and Kwiatkowksi, R. (1998, November). Emotional aspects of large group teaching. *Human Relations*, 51(11), 1403-1417.

Institute for Higher Education Policy. (2000). *Quality on the line: Benchmarks for Success in Internet Distance Education*. Washington, D.C.

Kirkman, B.L., Rosen, B., Gibson, C. B., Etsluk, P. E. and McPherson, S. (2002, August). Five challenges to virtual team success: Lessons from Sabre, Inc. *The Academy of Management Executive*, 16(3).

Kling, R. (2000). Learning about information technologies and social change: The contribution of social informatics. *Information Society*, 16(3), 217-232.

Maki, R. H., Maki, W. S., Patterson, M. and Whittaker, P. D. (2000). Evaluation of a Web-based introductory psychology course: Learning and satisfaction in on-line versus lecture courses. *Behavior Research Methods, Instruments and Computers*, 32, 230-239.

Moore, M. G. (1995). American distance education: A short literature review. In F. Lockwood (Ed.), *Open and Distance Learning Today*. New York: Routledge.

Neslar, M. S. and Hanner, M. B. (2001). Professional socialization of baccalaureate nursing students. *Journal of Nursing*.

Papert, S. (1980). *Mindstorms*. New York: Basic Books.

Russell, T. (2003). Retrieved March 30, 2003 from the World Wide Web: http://teleeducation.nb.ca/nosignificantdifference/.

Seigel, S. (1958). *Non-Parametric Statistics*. New York: McGraw-Hill.

Shipley, T. E. and Veroff, J. (1952). A projective measure of need affiliation. *Journal of Experimental Psychology*, 43, 349-356.

Svetcov, D. (2000). The virtual classroom vs. the real one. *Forbes*, 50-52.

Teleeducation (2003). Retrieved March 30, 2003 from the World Wide Web: http://teleeducation.nb.ca/nosignificantdifference/.

Tu, C. H. (2000). Online learning migration: From social learning theory to social presence theory in the CMC environment. *Journal of Network and Computer Applications*, 23, 27-37.

Chapter X

Modular Web-Based Teaching and Learning Environments as a Way to Improve E-Learning

Oliver Kamin
University of Göttingen, Germany

Svenja Hagenhoff
University of Göttingen, Germany

Abstract

This chapter can be assigned to the main fields of new and innovative educational paradigms and learning models, innovative modes of teaching and learning based on technological capabilities and strengths and weaknesses of technologies as effective teaching tools. It covers the construction of e-learning materials using a modular design approach in order to meet the technical and didactical requirements for the optimum operation of distance learning scenarios. First, it addresses the development path and substantial deficits of conventional e-learning materials. After this, it gives an overview of the requirements the supplier thinks necessary to develop high quality and state-of-the-art e-learning materials. In the following section, the customer's needs with regard to the e-learning materials will be addressed. Accommodating both parties and securing high quality requires a high flexibility for configuration of a Web-based learning and teaching environment. The next section introduces the respective concept based on modular structures. The content-related design of study modules will be shown with the support of an example taken from the education network WINFOLine.

Introduction

In e-learning, teaching materials can be accessed by a large and anonymous group of consumers via modern information technologies without temporal and local restrictions (Girmes, 1999). Successful implementation requires planning of the instruction units' technical and didactical guidelines to fit the learner's anthropogeneous and sociocultural conditions. A balanced combination of the selected contents, intentions, media and methods (Jank and Meyer, 2002) will achieve this. Many existing distance learning scenarios use e-learning pedagogies used for conventional presence teaching. Not only do these pedagogies not utilize the special advantages of the Internet as a distribution channel, they suffer two major deficits. First, conventional teaching materials are only suitable for small target groups and cover only a few learning types. Second, more conventional Computer-Based Trainings (CBT) and Web-Based Trainings (WBT) follow the construction Paradigms of Courseware Engineering. The core ideas of behaviorist teaching methods are hardly suitable for the self-controlled learning processes, because of their linear structure (Dichanz, 1994). These are the main reasons why the effectiveness and success of some traditional e-learning materials has been low (Gruber, Mandl and Renkel-Schwarzer, 2001).

Depiction and Evaluation of Conventional E-Learning Materials

Programmed Learning

Computer-based learning environments began in the early 1960s with "programmed learning." These environments focused on instilling factual knowledge with the help of programmed questions (such as fill-in-the-blank exercises and multiple-choice questions), testing, and other simple training exercises. They were based on a didactic design that relies on behaviorist principles. Through intermittent reinforcement and through the breaking up of content into very small units, the computer is to assume the role of the objective conveyor of knowledge (Schulmeister, 1997). The goal is to eliminate the teacher's mood changes and, thus, to be able to offer content of the same form and quality at all times. A major advantage of multiple and repeated access is assisting the individual process enabling students to more successfully memorize content than in the conventional classroom.

Empirical studies show that environments based on programmed learning have been introduced at schools. However, teachers have never really accepted them, because these teaching methods could not be integrated into conventional methods (Schulmeister, 1997). Due to the fact that, usually, only an individual adjustment of learning units and speed is possible, computer-based exercise programs have significant deficits in the areas of self-regulation, adaptation and individualization with regard to students (Tergan, 1992). The examination knowledge students have explicitly memorized through programmed teaching is also called torpid knowledge (Tramm, 1992). Even though this knowledge can be reproduced in the same context, students are neither able to transform it to similar scenarios, nor to apply it to an actual situation that calls for action.

Computer-Based Training

In the 80s and 90s, more complex and powerful hardware and software has transformed the computer-based learning into a CBT, which has, in addition to linear, also branched structure. In principle, the rigid structures of the programmed teaching software have been only partially eliminated or not eliminated at all in these systems (Schulmeister, 1997).

The main areas of usage for CBT are, once again, training and exercise situations. The teacher is replaced by the computer to spare him from routine tasks and to offer students the opportunity to access the learning materials at all times, as often as they want to. The teacher's responsibility has been changed from a mere conveyor of knowledge to a coach and motivator. The goal of conveying knowledge through CBT is to objectify the presentation of content and to make possible its multiple usages, even by other groups of students. Because of the lack of a conveyor of knowledge who is able to interact with the students, the computer is not able to understand the student's reactions and non-defined actions (Yass, 2000). Unlike programmed learning, CBT functions as multimedia tutorials and drill-and-practice learning systems, the latter having found wider realization (Sembill and Wolf, 1999). In addition to basic behaviorist principles, CBT includes elements of learning and cognitive psychology.

When the initial excitement wore off, CBT was not able to fully satisfy users because of its limitation to the conveyance of factual knowledge for reproduction only (Hundsinger, 2002). Additionally, the producer concentrated on a high programming quality, rather than on the realization of an ideal learning environment for self-directed and self-organized learning. One should keep in mind that CBT, constructed in the fashion we have described above, offers the student only a limited number of adaptation possibilities and, consequently, not much room for individualized learning options (Schulmeister, 1997). The program fails to recognize and evaluate the student's own ideas and numerous exercises can be solved by the trial-and-error system (Yass, 2000).

Web-Based Training

Conventional WBT environments mostly originate from CBT. In contrast to their predecessors, which were partially produced with the help of special software or were individually programmed, WBT is designed with the instruments and tools usually used for web sites and are thus accessible via network. The producer of WBT has new means of design through the net-like structure of teaching content which results from the usage of hypertext and hypermedia. In addition to that, the option of opening up new means of communication within the Internet and an intranet has liberated students from social isolation. At this point, however, many options are not fully used. Thanks to the fact that WBT was mostly based on CBT, most WBT makes use of the drill-and-practice system for the conveyance of knowledge. Consequently, WBT is only suitable for realizing less complex study objectives As opposed to CBT, conventional WBT has an added value for students. If the option is implemented in the respective WBT, the student is able to access additional or differently presented content at all times using the net-like structure. Practically speaking, though, conventional educational offerings tend to have a linear character despite these additional functions (Achtenhagen and Lempert, 2000) and do not use the potential of hypertext and hypermedia to a full extent. Therefore, this type of educational product is only of limited value for open learning, for multiple usage with various target groups and for teaching purposes (Dichanz, 1994).

In addition to that, conventional WBT does not satisfy the requirements that cognitivistic and constructivistic learning theories have on their respective environments. As a result, the WBT described above conveys mostly torpid knowledge due to its inherent behaviorist concepts. Therefore, knowledge is poorly transferable to real-life situations (Gruber, Mandl and Renkl-Schwarzer, 2001).

Design Requirements for E-Learning Materials

Preliminary Thoughts

When comparing the stage of development of already realized learning software with the requirements of the current teaching and learning research, one can see that the software does not meet the quality criteria and demands of e-learning. With the old software, the student cannot develop complex knowledge structures. Therefore, we should start to think about how the respective learning environments should be designed from the teacher's perspective (from now on also referred to as the supplier), in order to satisfy the aforementioned didactic needs and to facilitate the practical realization and operation of these arrangements. Both the supplier and the student (from now on also referred to as the consumer) are to be granted a high degree of freedom in order to realize their individual preferences.

E-learning products which are targeted at the asynchronous way (a time gap between creating and using information) of conveying information often show characteristics similar to conventional software. The CBT or the WBT mentioned in the introduction falls into this category, for instance. In particular, considerations for the roles of the products' suppliers and customers can be transferred to the e-learning software. Here the group of suppliers includes everybody who is concerned with content and technical development, the technical distribution, and the sale of the product. In addition to the actual consumer (e.g., the learner), the group of customers includes institutions that buy the pre-configured materials and make them available to their consumers (e.g., the students).

Suppliers' and Teachers' Requirements

The development of high quality learning materials is time consuming and requires large personnel expenditures (Hagenhoff, 2002). The materials should thus be used as frequently as possible. And instead of repeatedly employing them in one particular teaching session only, it is also conceivable that a product can be used in different sessions which are aimed at varying target groups and are thus connected to different teaching purposes. This includes, for instance, the usage in vocational schools, in in-house training or in-college education. One should keep in mind that the materials cannot only be used to support the acquisition of knowledge, but also can be used for research and evaluation of information in the area of knowledge management. This requires the e-learning materials to be adaptable to the preference structures of multiple supplier and customer groups with minimum effort. Instead of creating completely new learning environments for different target groups, the existing ones should be used again without creating additional costs. Therefore, the supplier aims at producing learning materials that can be realized easily, operated smoothly and maintained without problems. The possibility to import and export existing digital learning materials easily is of particular importance. We can achieve these objectives by producing modular e-learning materials. Standard tools and established methods of software development not only offer a maximum of functions, but the multimedia products produced with these tools can also be hosted on the same servers as conventional web pages. If an education supplier already owns a web server, the operation of the learning offerings causes little extra cost.

Customers' and Learners' Requirements

The asynchronous way of presenting and conveying information makes e-learning materials accessible at all times. With the help of the Internet and local intranets, the learner can access web-based learning environments easily. However, e-learning also has its drawbacks. The learner is prone to feel isolated because neither teachers, nor fellow students are around. Learning environments with a linear structure usually support only one learning type and the student is thus not able to leave the predefined studying path (Achtenhagen and Lempert, 2000). This is particularly problematic if teaching products are inappropriate for the target group. To keep the learner from being overstrained, we should structure the materials into easily comprehensible, self-contained segments and create a motivating atmosphere. We will clearly separate learning content and technological functions in order to improve the overview and to facilitate the configuration of the learning environment.

Construction Paradigms for Web-Based Teaching and Learning Environments

Hypertext and hypermedia are particularly useful to demonstrate the network structures of complex systems. Modern transmission protocols and standards ensure that the requirements for the construction of multimedia systems can be met. These technologies, as well as the modular conception of learning environments, help to meet the supplier's and customer's needs. We will call the modularly developed, e-learning materials Web-Based Teaching and Learning Environments (WTLE) in this chapter. The conceptual proximity to Complex Teaching and Learning Environments (CTLE) is intended because the research on CTLE is also applicable in the field of modern teaching and learning environments (Achtenhagen and John, 1992). In WTLE, the construction guidelines for cognitivistic and constructivistic learning theories are operationalized. We have to distinguish e-learning environments constructed according to these paradigms from conventional e-learning products (WBT and CBT), since the latter are based on construction principles like courseware engineering or programmed learning (Yass, 2000).

Modules and Interfaces

The WTLE elements can meet the needs of both teachers and learners by taking facets of learning theory into account. These modules, which are self-contained subsystems in terms of content or function (Göpfert, 1998), can be combined by using interfaces. On the one hand, these interfaces allow for system-specific connections, such as the use of common object libraries. These connections (dark lines in Figure 1) are made available through the framework module. On the other hand, we also need interfaces which provide content-specific connection structures (see the dark lines in Figure 2). With the help of these structures,

Figure 1.   Module Concept with System Specific Connection Structure

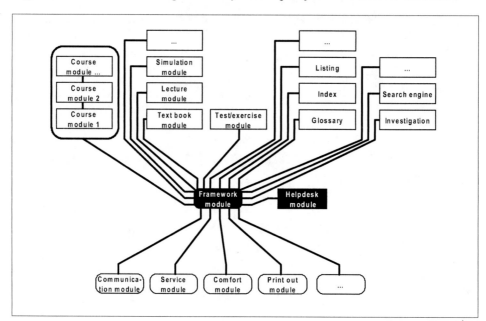

modules can be added and removed depending on suppliers' or customers' preferences. Generally speaking, adding and removing modules does not significantly influence the basic functions of other modules. The basic modules (colored dark in the middle of Figure 1) are an exception to this rule. Because they are required for the rudimentary operation of the WTLE, they cannot be removed.

System-Specific Modules

The system-specific modules (boxes with rounded edges in Figures 1 and 2) contain all functions related to the technical aspects of the learning environment. The learning system requires a comprehensive framework. As we can see in Figure 1, the framework module contains the global navigation for all parts of the WTLE. The centralized navigation structure allows all modules to access all parts of the system. This is the most important basic function of the learning environment. Content management systems can be used for this task. Besides, the framework module may enable the user to leave the predefined learning path and move freely within the WTLE. The communication module, which is also accessible via the framework module, offers different means of communication in order to accomplish synchronous and asynchronous tutoring scenarios between learner and tele-tutor. The service and comfort modules offer a wide variety of additional functions which are not required for the basic operation of the environment, e.g., systems check, individual configurations and storage of user data. The modular structure ensures that we can still use the WTLE after components are omitted or added. If additional modules, like communications or service, are not available, unrestricted access to the content of course modules is still possible. The remaining technical modules are configured on the now limited palette of functions of the arrangement.

Content-Specific Modules

Content-specific modules (boxes with pointed edges in Figures 1 and 2) contain the didactic elements of the WTLE. Special course modules allow the cognitivistic arrangement of learning content. In this context, each module has to represent a self-contained learning unit including statements about the unit's topic, the learning objectives, the required prior knowledge and the level of difficulty. With respect to the content, these modules can be linked arbitrarily. The possible connections are represented by the lines in Figure 2. Content-related structures will be realized by using the aforementioned learning objectives and definitions of prior knowledge, which serve as a content-specific interface.

A meaningful description of these parameters makes it easier for the future user of the teaching/learning environment to select modules and put them in order. Adding advanced study materials to the course modules (glossaries, examples for particular cases, etc.) is also possible with the help of the content-specific interface. With the help of meta-tags or similar descriptive elements within the objects themselves, we can gather information concerning the content-specific structuring and classification of the modules. Users can access these courses as a unit or on a by-module basis. Units are represented by transparent boxes with thick lines around the relevant module structures in Figures 1 and 2. As mentioned before, we can assign further modules or units to the existing individual modules or units as needed. Enrichments, which are particularly suitable in constructivistic learning environments, may be integrated by connecting modules like simulation, search or presentation, which support the acquisition and application of knowledge.

We can link the modules by defining learning objectives and prior knowledge. In addition to the module presenting the actual teaching subject, further content-specific modules are available. They contain a compressed overview of the subject represented by summaries, lists or glossaries. The learner is able to generate dynamic outputs and overviews using search and inquiry modules. In this case, information about the subjects treated in the modules serves as the necessary content-specific interface. We should be aware, though, that these

Figure 2.   Module Concept with Content Specific Connection Structure

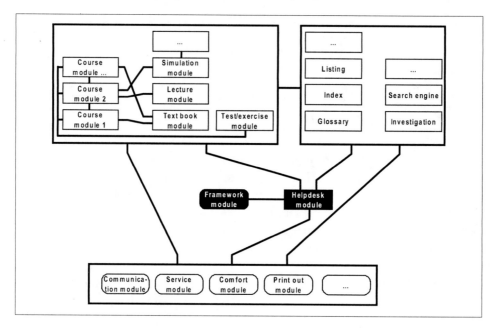

functions require a user who is familiar with the WTLE's controls. The helpdesk module gives the user detailed instructions for the operation of all parts of the WTLE and answers frequently asked questions.

Implementation Approach

As mentioned, the framework module provides the infrastructure for operation and use of the WTLE. Depending on the technology chosen for the transformation of the module concepts, the frame module may be a particular application which, with the help of established web-technologies, carries out the configuration of WTLE and provides the means for the user-specific presentation of content and functions. I will use the server-supported WTLE as an example for an implementing approach.

XML-documents describe all content- and system-specific modules of the server-supported WTLE — with the exception of the frame module and object libraries — in Figure 3. The linking of all XML-documents to one network or,

Figure 3.   Generating Process of a Server-Supported WTLE

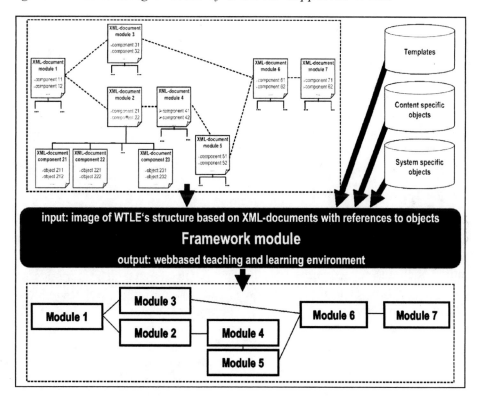

Figure 4.   Modules, Components and Objects

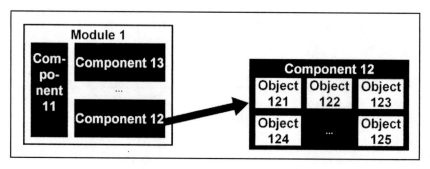

rather, to one structural concept, is a mirror image of the WTLE architecture as a whole. Each of the XML-documents contains all content- and system-specific information of a module and can thus refer to the relevant module components (see Figure 4) or navigate the entire WTLE (see Figure 1 and 2). After selecting the necessary components, the framework module, with the construction rules described in the XML-documents, automatically provides them. With regard to the server-based solution described in this section, one component is represented by one Web-page in a content-specific module. On this Web-page, a certain module displays various content and functions. The specific study material is, for example, represented by content-specific objects such as real-audio data, flash animations, graphics and text documents (see Figure 3). These objects, as well as the functions and procedures described in the system-specific modules, are stored in the respective libraries. Taken together, all of these components form the module (see Figure 4). By adding or leaving out components (here, Web-pages), an individual adjustment of the module can be achieved without seriously inhibiting the function of the other modules.

Phase-Oriented Approach

Conventional software engineering approaches can be used for the design of WTLE. We combine phase-oriented concepts with conventional curriculum planning methods to accommodate the special needs and characteristics of teaching and learning environments (Yass, 2000). According to Schumann, the development process can be divided into six phases (Schumann, 2003; Schach, 1990) or, as Yass suggests, into four steps: analysis, planning/specification, production and operation (Yass, 2000). The following paragraphs and Figure 5 demonstrate Schumann's six-phase concept and apply it to the specific needs of learning software development. Parallel to the software development process, a phase-comprehensive and continuing documentation process takes place. Authors, developers and operators of the educational product thus get sufficient information necessary to comprehend the individual steps at all times (Schumann, 2003).

In the initial proposal phase, we describe the relevant goals and content of the project in an abstract way. On this basis, the particular requirements are clarified in the specification phase. This phase then produces a specification document which is used as a basis for the design phase (Vision7, 2002) and contains the didactic concept of the WTLE. On the basis of this software specification document, the design guidelines describe the main functions and data of the educational product. In the design phase, we also produce scripts which outline the individual elements of the learning modules and bring content in the right order. In the final design specification phase, the design guidelines are specified with regard to relevant hardware and software problems (Schumann, 2003; Vision7, 2002). After completing the design guidelines and the final design specification, the production of WTLE takes place in the implementation or

Figure 5.   Phase-Oriented Concepts

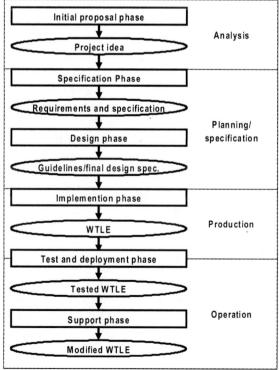

development phase (Schumann, 2003; Vision7, 2002). With regard to the modular products described here, the individual objects are produced first. In the following automatic configuration of the learning modules, it is of utmost importance that we bring module structure and content (here, objects and components) into a machine-readable format. While producing the learning module on a static HTML-basis, we initially generate web sites which contain references to the respective objects. Afterwards, we link these HTML-pages (components) according to the didactic concept (if necessary, with some additional functions). They form the complete education product.

During the test phase, the finished product has to pass comprehensive test runs and test scenarios to check if the requirements set in the software specification document and in the didactic concept are met. Stability and functionality of the education product are also tested. After eliminating all flaws, we make the WTLE available to the consumer or user (the student). The maintenance phase, which solves technical and content-related problems identified in practical usage, then completes the phase-oriented approach (Schumann, 2003; Vision7, 2002).

A Practical Example: Education Network WINFOLine

Configuration of the Online-Program Master of Science in Information Systems

We started the Education Network WINFOLine in 1997 to improve the course offerings at the participating universities in Göttingen, Kassel, Leipzig and Saarbrücken (Germany) with further e-learning classes. The number of classes in the department of Information Systems at the participating universities has been increased by up to 20%. Since 2002, WINFOLine features an internationally recognized master's degree. In this course of study, we use the e-learning materials of the four core universities, which have been tested and continually updated since 1998. In addition to that, we add e-learning materials from other partner universities. We have designed this as a correspondence course to meet the specific needs of (working) people who already have a college degree and want to receive the internationally recognized "Master of Science in Information Systems" as part of a work-and-study program. Students can access almost all courses via the Internet or other electronic means at all times and places.

Personal attendance at one of the four core universities is only required for a project seminar and for exams.

Figure 6 shows that the course of study is modularly designed. Each course usually consists of a WTLE and support services. We have divided the course of study into three majors: Information Systems, Computer Science and Business Management. The curriculum takes the prior knowledge of all students into account. For example, if a student has (some) basic knowledge of business management, he is not required to take (all) introductory courses in the business management basic study module. Instead, he can substitute them with advanced courses in Information Systems. Yet, we still want to make sure that the student can pass the study modules based on this (now skipped) basic module without problems. If the student has, for instance, put his main emphasis on Information Systems in his past studies (for example, in his initial studies, he took a class on databases), he may not, in his Master Studies, take an equivalent course in this subject. Instead, he has to look for an alternative course, such as Management Support Systems or Development of Application Systems (DAS).

When configuring further training measures, we need to take the prior knowledge of each individual student into account. This can be achieved with the help

Figure 6.   Curriculum WINFOLine Master Course of Studies

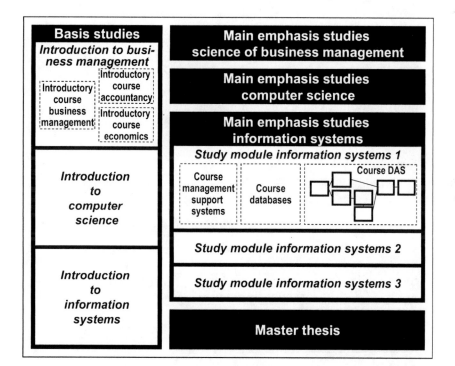

of the same description fields. We have analyzed this in the section before with regard to content and didactics. These description fields include information about content, learning objectives, prior knowledge, level of difficulty, and content and didactic-related structures of the objects in question.

We should also provide each course of studies with certain liberties for students to satisfy their individual needs. For the customer, the attractiveness and acceptance is thus raised and the product can succeed on the market. If we were to choose a highly restricted design, or rather, if our e-learning materials do not allow individual accommodations and combinations, this success would be in danger because redundant courses would be offered simultaneously and independent of each other.

We can plan and design the course offerings of the WINFOLine-Master-Degree by taking advantage of the same description fields already used in the WTLE-internal configuration. If we look at the whole course of studies or a package of several WTLE, we call this the macro-perspective. Micro-perspective, on the other hand, refers to the individual classes, which consist of a single WTLE, including the coaching. Table 1 shows the analogies between the systematic design of the WINFOLine Master course of studies (cf., Figure 6) and one individual WTLE. We can devise a fully automatic and individually shaped study plan if both levels are documented by identical description fields. An individually shaped presentation of the study materials follows. We can only achieve this because the configuration of the individual e-learning materials is based on each student's profile.

The Education Network WINFOLine will establish a further course of studies in the year 2003. The new course of studies has a different main emphasis and target group than the aforementioned Master's Degree. We want this new course of studies to attract students who don't have a German university degree as a pre-qualification. For this new and quantitatively more comprehensive course of studies, we will use the same pool of e-learning materials as a basis.

Table 1. Analogies Between the Systematic Design of the Winfoline master course of studies and a modular WTLE

Level of abstraction	Macro-perspective	Micro-perspective
Very high	Course of studies	-
High	Basis/main emphasis studies	Single WTLE
Middle	Study module	Module
Low	Course	Component
Very low	-	Object

Figure 7. Definition of a Content-Related WTLE-Module

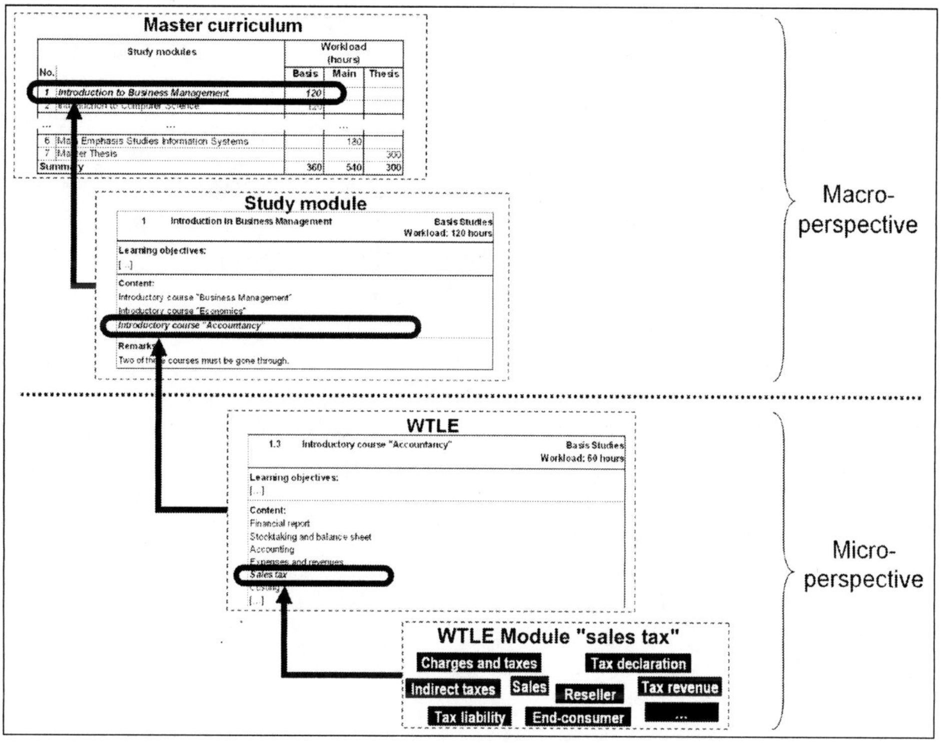

We can easily configure the existing course modules and tailor them to the special needs of the new target group using these particular description techniques. Furthermore, we will enrich the pool with other modularly designed e-learning materials which provide the basis for more WINFOLine courses.

Configuration of a WTLE for a Basic Course of the Online-Degree Master of Science in Information Systems

This practical example will demonstrate the design of the study module "Sales Tax" which is part of the WTLE "Introduction to Accountancy." Since the Fall 2002, WINFOLine have used this WTLE as an introductory course to the WINFOLine-Online-Degree "Master of Science in Information Systems." We have now entered the design phase of the WTLE. In order to develop the design guidelines of the WTLE or, rather, the individual study module, we first use the curriculum of the master's degree as a content-related foundation on a macro-perspective (see upper left side in Figures 6 and 7).

The program's design determines that the study module "Introduction to Business Management" consists of three different introductory courses students can choose from. The "Accountancy" course (framed on the second level in Figure 7) contains two components. The "Accountancy"-WTLE is one component, the other one holds a number of service options not further discussed here. On the micro-level, the WTLE-study module "Sales Tax" (framed on the third level in Figure 7) is formed. The dark boxes within the WTLE module on the fourth level in Figure 7 represent the individual components (Web-pages) with the study content. We have decided that, aside from integrating complex and problem-oriented exercises, we should take advantage of cognitivistic learning theory and add expressive graphics, animations and supplementary explanations at the end of each module. At some point, practical examples should support the content and students should be able to access them whenever necessary. By adding and removing components, the student can change the study module's level of difficulty. In this case, out of five possible levels of difficulty, a medium level is anticipated (see Table 2).

Table 2. Excerpt of a Script

Module 5: Sales Tax			
Comp. no.	Presentation sheet	Explanation text	Notes for the developing team
1	•Learning objectives: -The student should be able to represent important definitions in the area of sales taxes. (1) -[…] -The student should be able to execute all accounting activities which are concerned to the treatment of sales tax. (3) •Keywords: taxes, charges, sales, reseller, […] •Prior knowledge: accounting (3), expenses and revenues (4) •Level of difficulty: middle (3)		
…	…	…	…
3	Indirect taxes •The sales tax is a consumption tax. •It can be passed in a value-added-chain as follow: -From a company to another company, -to resellers -up to end-consumers. •At least the final consumers and not the resellers have to pay the sales tax.	The best known indirect tax is the sales tax. The operation of indirect taxes can be explained with the help of sales tax (tax debitors and tax carriers are not the same). The sales tax is a consumption tax, this means that the company will calculate the customer's taxable deliveries and will transfer the earned sales tax to the tax office. An added value arises on every layer of the chain. Every company have to pay only sales tax of the gained added value. At the sale of the supplier bills the respective buyer for the paid sales tax.[…]	•Visualize the system of indirect taxes with the help of a value-added-chain with Macromedia Flash. •Value-added-chain's elements has to display staggered. •Formulate suitable examples.
…	…	…	…

The study content of the "Sales Tax" module are characterized by various learning objectives (meaning, precise cognitive objectives). Because the WTLE functions as a content-related interface, we can make the respective connections. In this module, the student learns to understand the sales tax system and how to carry out the respective bookings. In order to realize this objective, we chose a cognitive learning objective on Level 3 (application), according to Bloom's taxonomy of learning objectives (Bloom, 1972). On the basis of the respective objectives and the content-related characteristics, we have defined the following precise objective:

The student should be able to execute all accounting activities which are concerned with the treatment of sales tax.

As we can see, a content (here, accounting activities and sales tax) and a behavior-related component (here, execute) form the learning objective. The learning objectives, prior knowledge and difficulty levels, keywords, as well as the specific content are summarized in scripts. In this context, we add meaningful numeric scales to the description fields in order to guarantee that they are machine-readable and can, thus, be easily evaluated.

Table 2 presents part of the script of the aforementioned study module. This segment contains two components. The script in this table is the result of the content-related concept. We use it as the main basis for further development of the WTLE and continually append it for documentation purposes. When developing the final design specifications, scripts are being analyzed with regard to a multitude of relevant technical software and hardware criteria. This specific example demonstrates that the animated graphic in component 3 is to be realized with the help of Macromedia Flash (see Table 2). In order to do so, we set the specific sizes, colors and times. The same is true for the explanations, where we arrange fonts and type size. These decisions are also important for the screen resolution for which this product is to be optimized. Now, we need a meaningful name for the objects and components in order to clearly identify them later. If we do not automatically generate components and modules with a special content management system, we need to clarify further questions for the application and for the necessary connections.

One has to decide and clarify, for instance, whether we want to realize the product with the help of static HTML pages or through a dynamic issuing of content with the help of a database. Even at this early point, we can see that we should take a multitude of design criteria into account in the final design specification document. For reasons of space, though, we will not discuss all of them in detail.

Figure 8.   Screenshot of Component 3 of the "Sales Tax" Module

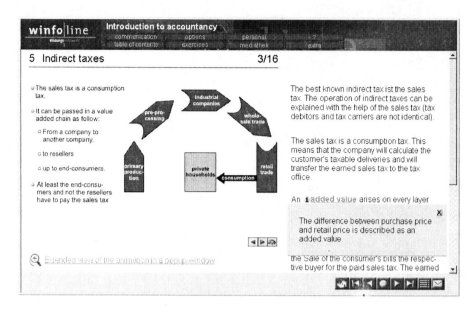

The following screenshots give the reader an impression of the WTLE described above. Figure 8 shows the realization of component 3, which is itself part of the "Sales Tax" module and has been described in Table 2. On the top and bottom of the screen, we can see the navigation options, which are made accessible through the framework module (see section, "Design Requirements"). According to the script (see Table 2), the flash-animation in the middle of the screen shows the value chain. On the left and right-hand side, we see text blocks which have also been specified in the script. In accordance with the script, the student finds further explanations and practical examples in the dark-colored box.

Figure 9 shows the WTLE-internal search engine, which also works offline, and belongs to the content-related modules. The search frame takes all modules into consideration. The white area on top represents the entry box with the results right underneath.

The comfort-module "Plug-in test," which is one of the system-related modules, is presented in Figure 10. This module checks whether the user's computer system has installed all the plug-ins or, rather, the programs necessary to use the WTLE. In case they are nonexistent, the user is asked to install them at this point. In this example, the student's computer system has passed the first three tests, which is symbolized by the green lights.

The last figure shows another content-related module. This time, it is a screenshot of a case study as a complex module, which is to convey additional knowledge to the learner in a constructivistic way. In accordance with Figure 2

Figure 9.   Screenshot of the Content-Related Module "Search Engine"

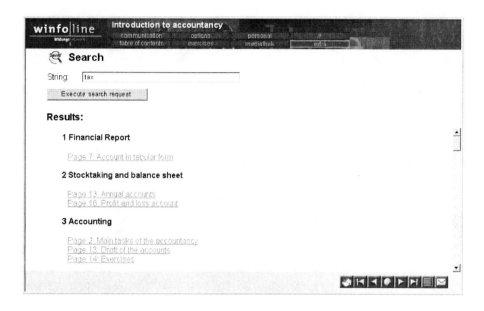

Figure 10.   Screenshot of the System-Related Module "Plug-in-Test"

Figure 11.   Screenshot of an Additional Lecture Module (Complex Case Study)

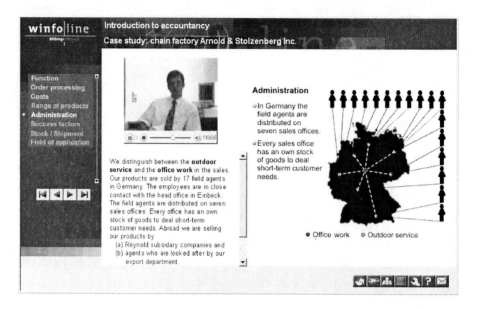

in the section before, an additional module helps to establish a content-related connection to the actual course. Based on the learner's preferences and technical prerequisites, the Realplayer's video presentation can be transferred into a audio or text presentation at all times. Multimedia elements can also be inserted or cut out. Even a combination of all of the different types of media is easily possible.

One can see that, when cutting out (with the exception of the frame module) or inserting additional modules, the operating condition of the "Accountancy" WTLE is still guaranteed. The user can also adjust it to his specific needs.

Summary

The development of high quality and economically sensible e-learning materials requires the consideration of various interests. In order to guarantee the multiple uses of study materials, it is wise to design and realize these products in a modular fashion. This is necessary to divide the immense developing costs between many customers in different fields of usage. By using already developed system components and resources, we can decrease the developing costs. Similar production methods have succeeded in the industrial production decades ago. Through a combination of e-learning software development, the insights from software engineering, and the product policy in industrial production, we have created new design opportunities to increase possible uses and, consequently, the potential for the success of e-learning in the education market.

Yet, we have to keep in mind that, when technically realizing modularly constructed education products, we have to fall back on already established standards and realization methods. Modular products depend on an open structure and possibilities for combination with other (possibly elsewhere developed) components. If we had chosen a highly restricted concept, this would lower — because of its lack of flexibility — the approval rate and the opportunities to use this product on the education market. When selecting the technologies to realize and distribute the materials, we should also keep in mind the aspect of multiple uses. So far, we were only able to use the implementation approach in server-based systems. Consequently, we had to exclude potential users who do not have Internet access. Therefore, we should add an offline option with similar configuration and distribution possibilities of a WTLE.

References

Achtenhagen, F. and John, E. (eds.). (1992). *Mehrdimensionale Lehr-Lern-Arrangements, Innovationen in der Kaufmännischen Aus- und Weiterbildung*. Wiesbaden: Gabler Verlag.

Achtenhagen, F. and Lempert, W. (eds.). (2000). *Lebenslanges Lernen im Beruf - Seine Grundlegung im Kindes und Jugendalter - Das Forschungs- und Reformprogramm* (Vol. 1). Opladen: Leske und Budrich Verlag.

Bloom, B. (1972). Taxonomie von Lernzielen im kognitiven Bereich, Basel.

Dichanz, H. (1994). Konzepte offenen Lernens - Chancen und Grenzen im Fernunterricht. In G. Zimmer (Ed.), *Vom Fernunterricht zum Open Distance Learning - Eine europäische Initiative*. Bielefeld: Bundesinstitut für Berufliche Bildung.

Girmes, R. (ed.). (1999). *Lehrdesign und Neue Medien: Analyse und Konstruktion*. München: Waxmann Verlag.

Göpfert, J. (1998). *Modulare Produktentwicklung — Zur gemeinsamen Gestaltung von Technik und Organisation*. Wiesbaden: Gabler Verlag.

Gruber, H., Mandl, H. and Renkel-Schwarzer, A. (2001). *Was lernen wir in Schule und Hochschule: Träges Wissen?* Retrieved May 17, 2002 from the World Wide Web: http://www.ph-weingarten.de/homepage/faecher/psychologie/konrad/handlungsorientierung/wissenHAN.htm.

Hagenhoff, S. (2002). *Universitäre Bildungskooperationen — Gestaltungsvarianten für Geschäftsmodelle.* Wiesbaden: Gabler Verlag.

Hundsinger, H. (2002). *Einsatz computerunterstützter Lernprogramme (CBT) in beruflichen Schulen — Erfahrungen aus einem BLK-Modellversuch.* Retrieved May 21, 2002 from the World Wide Web: http://www.leu.bw.schule.de/beruf/modvers/mv2/cbt-summary.htm.

Jank, W. and Meyer, H. (2002). *Didaktische Modelle.* Berlin: Cornelsen Verlag.

Schach, S. (1990). *Software Engineering.* Boston, MA: Aksen Associates Incorporated Publishers.

Schulmeister, R. (1997). *Grundlagen hypermedialer Lernsysteme, Theorie — Didaktik — Design.* München: Oldenbourg Verlag.

Schumann, M. (2003). *Entwicklung von Anwendungssystemen.* Retrieved March 18, 2003 from the World Wide Web: http://www.WINFOLine.uni-goettingen.de/wbt/eas.

Sembill, D. and Wolf, K. (1999). Einsatz interaktiver Medien in komplexen Lehr-Lern-Arrangements. Symposium Multimediales Lernen - Resultate und Perspektiven. In I. Gogolin and D. Lenzen (Eds.), *Medien-Generation. Beiträge zum 16.* Kongress der Deutschen Gesellschaft für Erziehungswissenschaft, Opladen.

Tergan, S. (1992). Wie geeignet sind computerbasierte Lernsysteme für das Offene Lernen in der beruflichen Weiterbildung? In F. Achtenhagen and E. John (Eds.), *Mehrdimensionale Lehr-Lern-Arrangements, Innovationen in der Kaufmännischen Aus- und Weiterbildung.* Wiesbaden: Gabler Verlag.

Tramm, T. (1992). Grundzüge des Göttinger Projekts Lernen, Denken, Handeln in komplexen ökonomischen Situationben- unter Nutzung neuer Technologien in der kaufmännischen Berufsausbildung. In F. Achtenhagen and E. John (Eds.), *Mehrdimensionale Lehr-Lern-Arrangements, Innovationen in der Kaufmännischen Aus- und Weiterbildung.* Wiesbaden: Gabler Verlag.

Vision7 (2002). *Software Development.* Retrieved March 24, 2002 from the World Wide Web: http://www.vision7.com/SoftwareDevelopment.

Yass, M. (2000). *Entwicklung multimedialer Anwendungen — Eine systematische Einführung.* Heidelberg: Dpunkt Verlag.

Chapter XI

The Effect of Culture on Email Use: Implications for Distance Learning

Jonathan Frank
Suffolk University, Boston, USA

Janet Toland
Victoria University of Wellington, New Zealand

Karen Schenk
K. D. Schenk and Associates Consulting, USA

Abstract

This chapter examines how students from different cultural backgrounds use email to communicate with other students and teachers. The South Pacific region, isolated, vast, and culturally diverse, was selected as an appropriate research environment in which to study the effect of cultural differences and educational technology on distance learning. The context of this research was two competing distance education institutions in Fiji, the University of the South Pacific and Central Queensland University. Three research questions were addressed: Does cultural background affect the extent to which students use email to communicate with educators and other students for academic and social reasons? Does cultural background affect the academic content of email messages? Does cultural background influence students' preference to ask questions or provide answers using email instead of face-to-face communication? To address these issues, two studies were conducted in parallel. Subjects were drawn from business information systems and computer information technology classes at the University of the South Pacific (USP) and Central Queensland University (CQU). Four hundred students at USP were surveyed about their email usage. In the CQU study, postings to course discussion lists by 867 students were analyzed. The results of these studies suggest that there are significant differences in the use of email by students from different cultural backgrounds.

Introduction

The impact of cultural diversity on group interactions through technology is an active research area. In common with the business world, the world of education is becoming increasingly globalized, and it is important for instructors to understand the impact of cultural differences on student learning. Cultural diversity represents an enormous challenge for global teams, but also offers a potential richness. Cultures and management styles often clash. People from different cultures have varying ideas about appropriate methods of communication and levels of accountability (Dubé and Paré, 2001). Acquiring the ability to manage cultural differences successfully can give a university significant advantages. The way forward is to foster cross-cultural learning and participation in education (Holden, 2002).

The following scenario highlights some cultural differences that may impact the distance learning process:

"Tenika, thank you for coming to see me. I'm concerned about your silence in class. Fifteen percent of your grade is based on class participation, and yet you never ask questions, express your opinions, or challenge the views of other students in class." Tenika Kepa, a Fijian M.B.A. student studying in Boston looked down, embarrassed. "Dr. Smith, I'm sorry, in my country to ask a question in class is considered rude. Fijians do not like confrontation. When our people disagree, they remain silent. This is often misinterpreted by people who demand or expect that we tell them to their faces what we do not want. We want them to be sensitive enough to feel that we don't agree. This is part of our culture." (Woodward, 2000)

Student engagement, discourse, and interaction are valued highly in "western" universities. Educators from individualist cultures like America, Europe and Australia may recognize Dr. Smith's frustration with Tenika's "quietness" in class. International students from Asian collectivist cultures may also empathize with Tenika's embarrassment at having to stand out against her group and express a personal opinion. With growing internationalization of western campuses, and rising distance learning enrollments, intercultural frictions are bound to increase.

There have been a number of papers that have examined the impact of cultural diversity and group interaction in computer-mediated communication environments (Jarvenpaa et al., 1998). This research adds to the body of knowledge by evaluating email effectiveness as a communications medium in facilitating meaningful class participation in two distance education institutions: The University of the South Pacific and Central Queensland University.

This chapter addresses three research questions:

1.   Does cultural background affect the extent to which students use email to communicate with educators and other students for academic and social reasons?

2.   Does cultural background affect the academic content of email messages?

3.   Does cultural background influence students' preference to ask questions or provide answers using email instead of face-to-face communication?

Literature Review

Hofstede's (1991) well-known model categorizes different cultures according to five pairs of dimensions (Figure 1).

Though his research has been criticized as being somewhat simplistic and dated, it provided a useful starting point for exploring possible differences between people from different cultural backgrounds (Holden, 2002; Myers and Tan, 2002). No formal research has yet been completed that maps Hofstede's model on the many South Pacific cultures, though there have been a number of recent publications reviewing aspects of the development of IT in the South Pacific (Davis et al., 2002; Olutimayin, 2002; Purcell and Toland, 2003). Lynch et al. (2002) have explored Hofstede's framework with respect to Fiji, hypothesizing where the indigenous Fijian population and the Indo Fijian population would fit on the framework. However, they are still in the process of collecting empirical evidence to validate their work. This research forms a useful starting point which makes it possible to locate South Pacific cultures on the dimensions of individualism, collectivism and power distance.

Figure 1.   Hofstede's Model of Cultural Differences

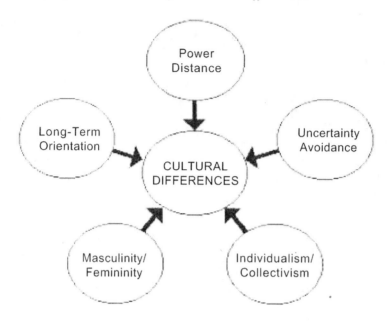

The literature has often cited difficulties in motivating students from Collectivist (as opposed to Individualistic) cultures to "speak up" in a face-to-face learning situation. Hofstede's evidence suggests that students from a collectivist culture prefer to listen, reflect, and discuss in person with peers, before preparing a written response. In many other collectivist cultures, it would be considered undesirable for students to speak up in class, as communication is mostly teacher-centered. In Fiji, lecturers have widely commented on the "quietness" of their students (Handel, 1998). Additionally, in some pacific cultural norms, student silence is seen as a sign of respect for teachers (Matthewson et al., 1998).

One problem with current research lies in interpretation. Most research has been carried out by westerners, who tend to judge students against their own cultural background (Jones, 1999). For example, some studies found that Chinese students from a collectivist culture adopt rote learning approaches, which were interpreted by western individualist teachers as inferior (Samuelowicz, 1987). More recent studies concluded that there is little evidence to support this belief and that Chinese students are no more likely to rely on rote learning than their Western peers (Watkins et al., 1991).

Most research on the effect of cultural differences has focused on traditional face-to-face teaching, rather than on distance education. More research is needed to understand fully the cultural contexts in which distance education programs are situated and how distance students process materials, especially in a second language (Guy, 1991).

A recent study investigated student media preferences from collectivist versus individualist cultures and concluded that collectivist students were significantly less comfortable with computer-mediated interaction, favoring face-to-face interaction, instead (Anakwe, 1999). The study concluded that students from collectivist cultures would be less receptive to distance education than students from individualistic cultures. It has been shown that students from individualistic cultures are also more willing to respond to ambiguous messages (Gudykunst, 1997), which may result in a different approach to email.

Other authors have suggested that a student's culture appears to influence online interactions with teachers and other students (Freedman and Liu, 1996); students from Asian and Western cultures have different web-based learning styles (Liang and McQueen, 1999); Scandinavian students demonstrate a more re-strained online presence compared to their more expressive American counter-parts (Bannon, 1995); and differences in participation levels were found across cultures in online compared to face-to-face discussions (Warschauer, 1996).

Cultural Differences

The Hofstede dimensions of interest for this research were individualism versus collectivism and high power distance versus low power distance. Hofstede's work indicated a strong relationship between a country's national wealth and the degree of individualism in its culture. Richer countries tend to have an individualistic style, whereas poorer countries are more collectivist. As a poor country grows richer it tends to move away from a collectivist pattern to an individualistic one. Additionally, people from a rural background tend to be more collectivist than those from an urban background. Therefore, identifying communication choice differences between students from urban and rural backgrounds was of interest to this study also.

A country that is collectivist is also likely to be a high power distance country, where the views of senior people tend not to be questioned. Pacific Island people fall into the high power distance with their system of chiefs and their tradition of not questioning the chief's decision. Society is also collectivist in nature with the custom of "Kere Kere" or not being able to refuse a favor that is asked of you by a member of your own in-group.

Indo-Fijian students come from a somewhat different background. Indo-Fijians have lived in Fiji for at least three generations and because they arrived as immigrants or through the Girmit (bought in labor) system, they appear not to have the strong family ties that native Fijians have. Previous work has classified India as a high power distance, collectivist country; however, India scores much higher on the individualism index than most other poor countries. The caste system among the Indians within Fiji is not as prominent as in India; therefore, power distance tends to be smaller than among the Fijian population. Indo-Fijians have lower uncertainty avoidance and are more inclined to take risks, which helps to explain their success in the commercial sector in Fiji. Therefore, although both groups of students are collectivists, the Pacific Island students would be expected to be more strongly collectivist than the Indo-Fijian students. It should be noted that these classifications are very broad, as Lynch et al. (2002) note there are significant divisions between Indo-Fijian groups, based on caste and religion and also between indigenous Fijians based on different regional and historical affiliations. Within these two main and quite distinct cultures, each individual will construct their own value system depending on their cultural background and life experiences. However, the Hofstede framework makes a useful starting point for beginning to understand broad cultural differences.

In conflict situations, members of collectivist cultures are likely to use avoidance, intermediaries, or other face-saving techniques. In collectivist societies, in-group/out-group distinctions continue in education settings so that students from different ethnic or clan backgrounds often form sub-groups in class. They work together on assignments and there is often little interaction with students from different sub-groups.

In a distance education context this can make the summer school teaching experience in an island center very different from main campus teaching, as the summer school students are all part of the same in-group. Within the region, variations in cultural attitudes towards technology have been noted by various visiting lecturers. Two recent surveys of information distribution and communication technology showed marked variations in computer and Internet access among different countries of the Pacific Island region (Landbeck, 2000; Lockwood, 2000).

The next part of this section will focus on understanding and modeling the differences between the University of the South Pacific (USP) and Central Queensland University (CQU).

Knowledge Transfer

The two institutions studied have different approaches to education. The University of the South Pacific (USP) is owned and operated by 12 Pacific Island countries. USP serves a population base of more than 1.8 million people. The member countries are scattered more than 32 million square kilometers of ocean. Within the South Pacific region there is great cultural, language, and economic diversity, with more than 200 languages spoken (Landbeck and Mugler, 2000). USP was established in 1966 as a dual mode institution to offer both campus and distance learning courses. In 2000 student enrolment reached 9,118 (46% external). Similar to many universities around the world, USP has experienced rapid growth, with first year enrollments in mathematics and computing science courses rising as high as 700 students (Daniel, 1996).

Central Queensland University, based in Rockhampton, Australia, has recently opened a campus in Suva, Fiji. The campus has about 1,000 students from China, Korea, India, Bangladesh and Fiji. CQU's public relations office has extended more than 2,000 offers to students for the 2001 school year. CQU Fiji has an electronic library that links to the CQU main library in Rockhampton. Every student has access to email and the Internet. The university has a policy of interactive, system-wide learning.

For the purpose of model building, USP can be viewed as a traditional university where knowledge transfer tends to occur in a one-directional mode. In many developing world universities, it is traditional for educators to lecture and students to listen and learn. Publicly questioning the teacher is considered rude. Knowledge flows from the "professor" or "lecturer" to the student. This familiar model is often called the *container* or *transfer* model of knowledge transfer, or *migratory* knowledge (Badaracco, 1991). An example of a "pure" container model might be the traditional distance learning via a correspondence course. USP's distance learning model is somewhat similar in that learning materials are packaged for students and little interaction is anticipated between student and teacher. Figure 2 proposes an adaptation of the container model incorporating a bridging function (Jin et al., 1998), as well as components to reflect the affect of distance learning technology and cultural factors on knowledge transfer.

In contrast, Central Queensland University course pedagogy is extraordinarily dependent on email communication. Fifty percent of students' grades are based on group exercises. Groups consist of five to 10 students from 12 countries. Students are assigned to groups by the course coordinator to maximize diversity. Students are required to post within group and between group evaluations to a listserv each week (Romm, 2001; Jones, 1999). Its teaching philosophy empha-

Figure 2. Distance Learning Model with One-Way Knowledge Transfer

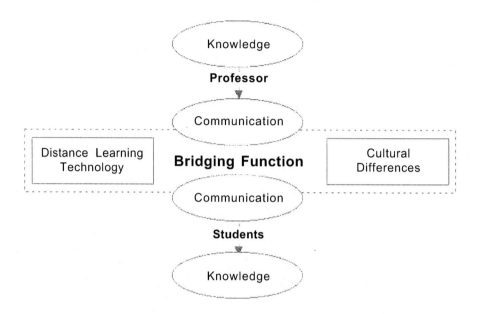

sizes the importance of student-student and student-teacher interaction using computer-mediated communication. A large percentage of a distance learning student's grade is based on teamwork. Students are encouraged to learn from each other, as well as from the teacher.

The *social construction* model of learning and knowledge transfer was deemed appropriate for CQU. This model represents knowledge as one part of a process. It considers knowledge, cognition, action, and communication as inseparable. "The term *enactment* captures this interrelationship among the different aspects of knowing, acting, communicating, and perceiving. Knowledge takes on meaning as the entity interacts with its environment through communicating with other entities, acting (and, thereby, changing the environment), and interpreting cues arising from these interactions" (Weick, 1979).

Figure 3 modifies an extension of Weick's work (Jin et al., 1998) to incorporate technological and cultural factors that might affect the success of a distance education "social constructivist" view. As students in virtual teams perceive that communication often is imperfect even when the students agree on a course of action. The problems associated with flawed communication often result in a knowledge transfer breakdown. Badaracco (1991, p. 82) notes this when he refers to "*embedded knowledge*, knowledge that resides in the relationships between and among individuals and groups."

Figure 3.   Distance Learning Model with Socially Constructed Knowledge

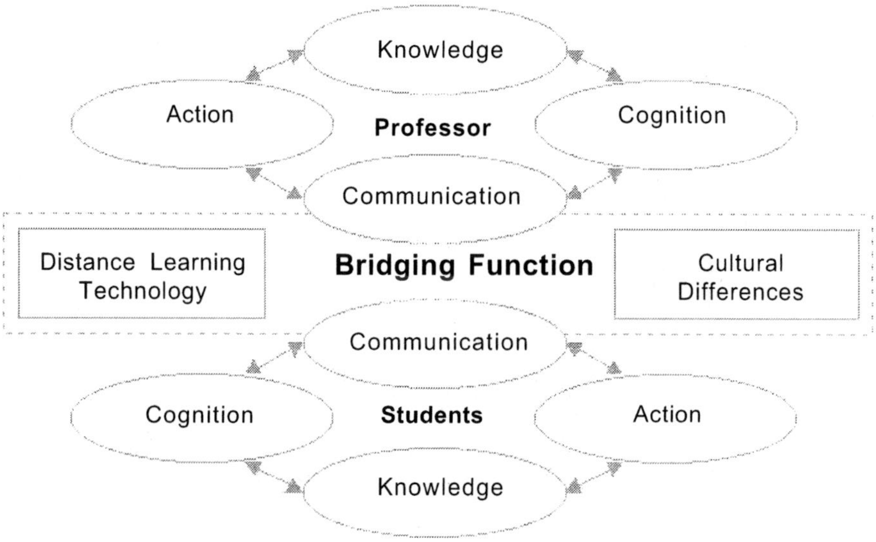

The University of the South Pacific (USP) Study

In 2001/2002 a questionnaire survey was conducted at USP to ascertain students' use of such facilities as email and Internet and to look at the differences in provision between rural and urban areas. A sample of almost 400 undergraduate students was surveyed. The majority were Fiji based students studying on the main campus by face-to-face mode.

Seventy-four percent of the sample were full-time rather than part-time students. The majority (53%) of respondents were aged between 20 and 24 and approximately two thirds (66%) were male. The students were taking a range of subjects, but slightly over half the sample were studying for Computer Science and/or Information Systems majors. The majority of the sample (65%) were Fijian or Indo-Fijian, however, some data was collected from summer schools conducted in Kiribati, Vanuatu and the Solomon Islands, which gave a more representative sample of the USP students. The summer schools were also taught in a face-to-face mode for intensive six-week blocks. See Figure 4 for sample breakdown by country.

The questionnaire was anonymous, and due to ethnic tensions in Fiji at the time of the survey, students were not asked to identify their nationality. Therefore, no

Figure 4.   Number of Survey Respondents by Country

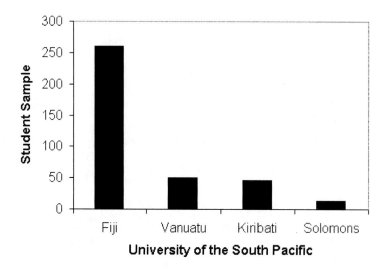

University of the South Pacific

distinction can be made between Fijian and Indo-Fijian students in terms of cultural background. From observation, roughly 75% of the students taking Computer Science and Information Systems in Fiji were Indo-Fijian. There were no significant numbers of other ethnic groups in the data from the regional centers.

Results and Discussion

Approximately 90% of the students reported having email access that was used mainly for social purposes. More than 65% of students used email more than 10 times per semester to contact friends (see Table 1).

Is email used to a more limited extent for purposes of teaching and learning? Less than 2% of students surveyed would choose email as their first choice method to ask a lecturer a question about a course. As illustrated in Table 2 below, emailing a lecturer to ask a question is the fourth most frequent choice. The majority of students prefer to consult the lecturer face-to-face or to ask a fellow student. However, it should be noted that students do not feel that they are at a disadvantage by choosing not to use email. Most (75%) of the students reported being able to get answers for questions they put to lecturers within a maximum

Table 1.   Breakdown of Numbers of Email Messages Sent to Friends by Students per Semester (for Whole Student Group)

		Frequency	Percent
	Never	65	16.4
	One to five per semester	40	10.1
Valid	Six to 10 per semester	28	7.1
	> 10 per semester	254	64.0
	Total	387	97.5
Missing		10	2.5
Total		397	100.0

of two days. The results indicate that many USP students have no real need to use email as their traditional methods of asking friends or asking the lecturer face-to-face have proven effective.

About half the students surveyed (46%) never use email to contact their lecturers and the remainder use it infrequently. Ninety percent of students send five or fewer messages a semester for this purpose.

Table 2.   Ranking of Email as a Method of Asking a Lecturer a Question, Out of a Choice of Five Methods (All Students Included)

		Frequency	Percent
	1	6	1.5
	2	37	9.3
	3	90	22.7
Rank	4	171	43.1
	5	89	22.4
	Total	393	99.0
Missing		4	1.0
Total		397	100.0

Table 3.   Numbers of Emails Sent to Students by Lecturers (All Students Included)

		Frequency	Percent
Valid	Never	124	31.2
	One to five semester	97	24.4
	Five to 10 semester	80	20.2
	More than 10	94	23.7
	Total	395	99.5
Missing		2	.5
Total		397	100.0

Approximately one third of students never received email from lecturers and 50% either received no email at all or fewer than five messages per semester. (See Table 3.) Several students commented that they received email from their information systems and computer science lecturers, but not from lecturers in other subjects: information systems and computer science students comprised the largest percentage of students who received 10 or more messages per semester from their lecturers.

The majority of the CS and IS courses use web pages stored on the university intranet to hold information such as lecture notes and assignment details. The main teaching model used is a one-way flow of information from teacher to student with a very limited use of two-way communication. Students were also less likely to use email to contact the administration, with the majority (88.6%) indicating that they never emailed the administration.

Students did take advantage of the opportunity to access the Internet and the USP home page. Sixty percent of students reported that they accessed the Internet and the USP homepage at least once or twice per semester. However, students were unlikely to use the online journals available through the library. Only about a quarter of students used this facility.

When the data from other South Pacific regions are compared with Fiji, a clear contrast emerges. Other regions show much less use of email. (See Figures 5 and 6.) In many cases, regional students had as much opportunity to use email as main campus students. For example, in Vanuatu there is a student:PC ratio of 10:1, as compared to 30:1 in Fiji, so lack of access to technology alone does not explain the differences seen.

Figure 5. Emails Sent to Students by Lecturers: Fiji vs. Rest of South Pacific (All Students Included)

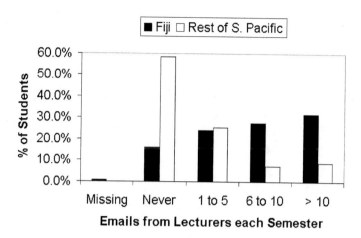

There seems to be a significant difference between emails received from lecturers by rural versus urban students. A rural student appears to be less likely to receive an email from their lecturer than an urban student. The results for both emails sent to lecturers and friends were not significant. Further examination indicates that neither rural nor urban students are likely to send emails to lecturers. The use of email for teaching and learning appears to be low (Table 4).

Figure 6. Emails Sent by Students to Lecturers: Fiji vs. Rest of South Pacific

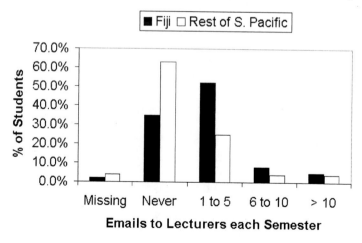

Table 4. Emails Sent to Lecturers by Students (All Students Except Kiribati)

Student Home	Never	One to five per semester	Five to 10 per semester	More than 10	Total
Rural	10.6%	12.1%	1.8%	.9%	25.3%
Town	16.5%	17.9%	2.4%	1.8%	38.5%
City	12.4%	19.1%	2.4%	2.4%	36.2%
Total	39.4%	49.1%	6.5%	5.0%	100.0%

The number of emails sent to friends was analyzed. No significant relationship was found among rural, town and urban students regarding frequency of email use to contact friends (see Table 5). These results indicate that there is no difference between students of different collectivist cultural backgrounds in their frequency of email use for social purposes.

To examine more fully the rural versus urban dimension in regards to email usage for teaching and learning, it was decided to examine the data set of main campus students. All main campus students should have similar access to technology. Therefore, a comparison of students from a rural and an urban background would not be influenced by such issues as less sophisticated technology or a lack of access to technology.

When the email use by main campus students from rural and urban backgrounds was compared using Pearson chi-square, no significant relationships were found. These results tend to suggest that when students have an equal access to technology, their use of email will be similar independent of the extent of their collectivist cultural background.

Table 5. Count of Emails Sent by Students to Friends (All Students Except Kiribati)

	Never	One to five per semester	Six to 10 per semester	More than 10 per semester	Total
Rural	2.4%	2.7%	1.5%	18.6%	25.1%
Town	3.8%	3.8%	3.8%	27.2%	38.8%
City	.9%	3.3%	3.0%	29.0%	36.1%
Total	7.1%	9.8%	8.3%	74.9%	100

The Central Queensland University Study

Central Queensland University makes use of a list server in all distance classes. Membership to the lists is open to the public and students are warned that all their posts to the list are accessible to the public.   Two semesters of student posts were analyzed covering the period: July 16, 2001 to December 14, 2001. As the University of South Pacific survey had focused on student use of email in information technology classes, a similar data set was chosen for CQU.   The Faculty of Informatics and Communication at http://www.infocom.cqu.edu.au/ Courses/ delivers business oriented and technology oriented computer classes. This study focused on 21 courses in these two areas.

Names and email addresses of more than 1,500 individuals who had posted messages to the discussion lists were recorded. Students had email domain names from Australia, Brunei, Fiji, Hong Kong, Malaysia, New Guinea, Singapore, and Taiwan. The first step was to extract students from Australia and Fiji with an email domain of .fj or .au. Students with email addresses from Hotmail, Yahoo, etc., were not included. Several students participated in more than one class, but most names were unique.

The next step was to separate the teachers from the students. It was assumed that teachers and teaching assistants would be regular contributors to their class discussion list. After tabulating the number of posts by each remaining participant a small group of very frequent posters was investigated. This involved checking each class syllabus for teachers and assistants names. A further check for frequent poster email addresses on Google determined whether these participants were affiliated professionally with CQU. This led to the removal of 42 individuals from the sample.

The final step was the most difficult and subjective taken to refine the sample. The study's purpose was to compare students from collectivist and individualist cultures. The authors decided, therefore, to remove "western" students from the Fiji sample and to ensure that the Australian sample included only "western" students. In Fiji there were only a few students with "western" names who were removed. In Australia about 15% of students had names that appeared to several investigators to be Asian in origin. It was impossible to determine whether these students had just arrived in Australia, or had lived there for generations. However, to avoid the possibility of confounding study findings with a mixed individualist/collectivist Australian sample, the individuals identified were excluded.

The remaining sample consisted of 653 students from Australia studying information technology classes and 160 studying business information systems classes. From Fiji, there were 47 students studying information technology classes and seven studying business information systems classes. Given the small sample size of Fijian students, the sample of business IS classes and computer IT classes were combined.

Results and Discussion

As expected, Australian students send significantly more posts to their discussion board than Fijian students. This confirms the hypothesis that Australian students (individualist culture) would be more active in discussion lists than Fijian students (collectivist culture).

The final stage of this analysis focused on the content of messages posted to the discussion lists. A random sample of 260 messages was drawn from Australian and Fijian students. Message content was coded as either a question, an answer, or social. Analysis indicates that there are significant differences between Fijian and Australian students' posting behavior (Table 6).

Australian students appear more ready to respond to questions than Fijian students. Fijian students volunteered fewer answers to the list. One plausible explanation might be that Fijian students feel anxious about "losing face" among their peer group if their answer is inappropriate. Another possibility could be that Fijian students view participation on the list as not directly affecting their grade and, therefore, see no reason to volunteer answers. An analysis of Fijian messages confirmed that a large percentage of messages were social in nature. Students appear to use the lists for forming groups more often than Australian students. Further investigation of questions asked by Fijian students indicated a need for reduction in ambiguity about assignment specifications. Analysis shows

Table 6. Comparison of List Posting Behavior by Message Type

	Fiji Students	Australian Students
Question	52	40
Answer	36	77
Social	35	20
	CHI-SQUARED	19.83
	SIG	0.00005

that Australian students had proportionately fewer questions, which were primarily technical in nature, or, interestingly, focused on what courses they should take next semester.

Conclusion

This chapter examined how students from different cultural backgrounds use email to communicate with other students and teachers. The South Pacific region, isolated, vast, and culturally diverse, was selected as an appropriate research environment in which to study the effect of cultural differences and educational technology on distance learning. The context of this research was two competing distance education institutions in Fiji, the University of the South Pacific and Central Queensland University. Email was used for teaching and learning in different ways at these institutions. At Central Queensland University, email was incorporated into teaching pedagogy across all courses, whereas at the University of the South Pacific email was available to students, but usage was voluntary.

Three research questions were addressed: Does cultural background affect the extent to which students use email to communicate with educators and other students for academic and social reasons? Does cultural background affect the academic content of email messages? Does cultural background influence students' preference to ask questions or provide answers using email instead of face-to-face communication? To address these issues, two studies were conducted in parallel. Subjects were drawn from business information systems and computer information technology classes at the University of the South Pacific and Central Queensland University. Four hundred students from the University of the South Pacific were located at different regional centers were surveyed about their email usage. In the Central Queensland University study, postings to course discussion lists by 867 students based in Fiji and Australia were analyzed. The results of these studies suggest that there are significant differences in the use of email by students from different cultural backgrounds and that cultural differences do impact student preferences and practices with regard to email use.

The USP study suggested that collectivist students are more likely to use email to interact socially with their peers than they are to use it for contacting their lecturers. The USP study compares the numbers of emails sent by students from the regions with students from the main centers. There is evidence of less email usage by students from a rural background, but this result is not conclusive. The influence of other factors such as unequal access to technology could not be ruled out.

The CQU analysis found evidence of a greater usage of email by students from an individualistic culture, as opposed to students from a collectivist culture. Collectivist students appear to send different types of messages. They tend to ask more questions than individualist students and their questions are more likely to focus on group formation and clarification of assignment ambiguity. The individualistic students are more likely to volunteer answers than the collectivist students. Anxiety over appropriateness of one's answer may contribute to collectivist students' reluctance to volunteer answers.

As both studies were of a limited sample of individualist/collectivist students in the Pacific region, generalization of these findings should be treated with caution.

This quantitative survey suggests reluctance by students from a collectivist background to use email for teaching and learning, although there is a widespread social use of email. A useful follow up would be to carry out qualitative research, interviewing students and academics to gain a more in-depth understanding of the reasons for some of the observed behaviors. Additional research in this area is important as we expand teaching across cultural boundaries through the use of distance learning.

References

Anakwe, U. et al. (1999). Distance learning and cultural diversity: Potential users' perspective. *International Journal Of Organizational Analysis*, 7(3), 224-244.

Badaracco, J. L., Jr. (1991). *The Knowledge Link: How Firms Compete through Strategic Alliances*. Boston, MA: Harvard Business School Press.

Bannon, L. J. (1995). Issues in computer-supported collaborative learning. In C. O'Malley (Ed.), *Computer Supported Collaborative Learning* (pp. 267-281). Berlin: Springer-Verlag.

Daniel, J. S. (1996). *Mega-Universities and Knowledge Media: Technology Strategies for Higher Education*. London: Kogan Page.

Davis, C. H., McMaster, J. and Nowak, J. (2002). IT-enabled services as development drivers in low-income countries: The case of Fiji. *The Electronic Journal on Information Systems in Developing Countries*, 9(4), 1-18.

Dubé, L. and Paré, G. (2001). Global virtual teams. *Communications of the ACM*, 44(12), 71-73.

Freedman, K. and Liu, M. (1996). The importance of computer experience, learning processes, and communication patterns in multicultural networking. *Educational Technology Research and Development*, 44(1), 43-59.

Gudykunst W. B. (1997). Cultural variability in communication. *Communication Research*, 24(4), 327-348.

Guy, R. K. (1991). Distance education and the developing world: Colonisation, collaboration and control. In T. D. Evans and B. King (Eds.), *Beyond the Text: Contemporary Writing on Distance Education* (pp. 152-175). Geelong, Australia: Deakin University Press.

Handel, J. (1998). Hints for teachers. *Centre for the Enhancement of Teaching and Learning*, University of the South Pacific.

Hofstede, G. (1994). *Cultures and Organisations*. New York: Harper Collins.

Holden, N. J. (2002). *Cross-Cultural Management: A Knowledge Management Perspective*. London: Prentice Hall.

Jarvenpaa, S. L. and Leidner, D. (1998, June). Communication and trust in global virtual teams. *Journal of Computer Mediated Communication*, 3(4).

Jin, Z. et al. (1998). Bridging US-China cross-cultural differences using Internet and groupware technologies. *The 7th International Association for Management of Technology Annual Conference*. Retrieved from the World Wide Web: http://www.cim-oem.com/bridge_8c18c.html.

Jones, A. (1999). The Asian Learner — An Overview of Approaches to Learning. Working Paper Series, *Teaching and Learning*, Unit Faculty of Economics and Commerce, University of Melbourne. Retrieved from the World Wide Web: http://www.ecom.unimelb.edu.au/tluwww/workingPapers/theAsianLearner.htm.

Jones, D. (1996). Solving some problems of university education: A case study. In R. Debreceny and A. Ellis (Eds.), *Proceedings of AusWeb '96* (pp. 243-252). Lismore, NSW: Norsearch.

Jones, D. (1999). Solving some problems with university education: Part II. *AusWeb99, the 5th Australian World Wide Web Conference*. Retrieved from the World Wide Web: http://ausweb.scu.edu.au/aw99.

Landbeck, R. et al. (2000). Distance learners of the South Pacific: Study strategies, learning conditions, and consequences for course design. *Journal of Distance Education*, 15, 1.

Liang, A. and McQueen, R. J. (1999). Computer assisted adult interactive learning in a multi-cultural environment. *Adult Learning*, 11(1), 26-29.

Lockwood, F. et al. (2000, August). Review of Distance and Flexible Learning at the University of the South Pacific. *Report submitted to the Vice Chancellor and senior staff at the University of the South Pacific*.

Lynch, T., Szorengi, N. and Lodhia, S. (2002). Adoption of information technologies in Fiji: Issues in the study of cultural influences on information technology adoption. *Conference on Information Technology in Regional areas*, Rockhampton, Australia.

Materi, R. A. (2000). The African virtual university — An overview. *Ingenia Training*.

Matthewson, C. et al. (1998). Designing the rebbelib: Staff development in a Pacific multicultural environment. In C. Latchem and F. Lockwood (Eds.), *Staff Development in Open and Flexible Learning*. London: Routledge.

Myers, M. and Tan, F. (2002). Beyond models of national culture in information systems research. *Journal of Global Information Management*, 10(1), 24-32.

Olutimayin, J. (2002). Adopting modern information technology in the South Pacific: A process of development, preservation, or underdevelopment of the culture. *Electronic Journal of Information Systems in Developing Countries*, 9(3), 1-12.

Pacific Islands Development Program/East-West Center (2001). Fiji Battling Education War At Tertiary Level. Retrieved from the World Wide Web: http://166.122.164.43/archive/2001/February/02-19-11.htm.

Purcell, F. and Toland, J. (2004). Electronic commerce for the South Pacific: A review of e-readiness. *Electronic Commerce Research Journal, 4*(3) (Forthcoming).

Romm, C. et al. (2001). Searching for a "killer application" for on-line teaching — Or are we? *Proceedings of AMCIS2001*, Boston, Massachusetts, USA.

Samuelowicz, K. (1987). Learning problems of overseas students: Two sides of a story. *Higher Education Research and Development*, 6, 121-134.

Tan, B. C. Y., Wei, K. K., Watson, R. T. and Walczuch, R. M. (1998). Reducing status effects with computer-mediated communication: Evidence from two distinct national cultures. *Journal of Management Information Systems*, 15, 1.

Warschauer, M. (1996). Comparing face-to-face and electronic discussion in the second language classroom. *CALICO Journal*, 13(2), 7-26.

Watkins, D. et al. (1991). The Asian learner as a rote learner stereotype: Myth or reality? *Educational Psychology*, 11(1), 21-34.

Weick, K. E. (1979). *The Social Psychology of Organizing* (2nd Edition). New York: McGraw-Hill.

Woodward, T. (2000). Paraphrased from a speech by Taina Woodward, a Fijian citizen, presented at the Beijing + 5 Conference, United Nations, New York City, 8 June 2000.

Section III

Building an Organization for Successful Distance Educations Programs

Chapter XII

A Strategy to Expand the University Education Paradigm

Richard Ryan
University of Oklahoma, USA

Abstract

To date most online content and experiences have been packaged in a traditional "class" format and delivered using a web site posted on a provider's server. This chapter suggests a slight deviation from this approach for packaging and delivering Internet education. It suggests a look beyond the "class" delivery approach. The premise for this strategy is the belief that the greatest strength of the Internet for education may lie in delivery of class "components," not classes, themselves. These online components can be used to supplement and add value to the traditional class experience, not replace it. The strategy proposes that the university provide, sponsor, administer and maintain an automated online portal to post and sell faculty-created material. An "e-store" selling products developed by the university's faculty members. It is hoped that universities will explore this idea to develop new ways of packaging and delivering education that better reward the faculty developer, help pay for the service and also add "value" to the education experience.

The Internet is maturing as a delivery medium for university education. The previous phase entailed developing a presence. It was the push to "have a web site." It is hoped that we have moved beyond this necessary step into the next phase. It is also hoped that this current phase will bring exploration of ways for university education developers and users to harness and optimize the emerging capabilities of the Internet to add "value" to the education experience. New ways of packaging, selling and delivering education hopefully will be explored. "Online instruction can offer new challenges and opportunities to both students and instructors. Most students do not view online instruction as replacement for traditional classroom instruction. However, with the right subject matter, with the right instructor or facilitator, and for the right student, Internet or online classes can provide an effective educational environment and offer a viable alternative to traditional classroom instruction" (Cooper, 2001).

Today there are great opportunities for creating enriched classes combining traditional and online resources. Traditional methods of presenting information in the classroom are very limited when compared to what can be done today in an online interactive, automated format using the Internet. The potential for online experiences to supplement, replace or exceed traditional classroom experiences becomes more apparent as technology advances and new efforts are placed online. Advanced web development and online database capabilities combined with adequate bandwidth, provide greater opportunity to optimize interactive Internet capabilities. Instructional design of online learning and assessment can incorporate database capabilities to deliver and administer information or exercises, automatically assess performance, track the results and provide immediate feedback to the user and/or Instructor. This online model is very efficient for performance-based learning and assessment.

To date most online content and experiences have been packaged in a traditional "class" format and delivered using a web site posted on a provider's server. Server space, the input format and operation of the web site, is provided typically by a university or commercial portal. Purchase (if required), access and delivery are traditional, except for the anywhere/anytime advantage of the Internet. In many cases universities provide this service to their students using a courseware "shell."

This chapter suggests a slight deviation from this approach for packaging and delivering Internet education. The idea represents a shift in the existing online education paradigm. Instead of just substantiating "why we should shift," this chapter suggests a model to look beyond the "class" delivery approach. The premise for this strategy is the belief that the greatest strength of the Internet for education may lie in delivery of class "components," not classes. The online components can be used to supplement and add value to the traditional class experience, not replace it. The Internet provides new opportunities to do things that can't be done in the traditional classroom setting. It provides access to resources that otherwise could not be explored and used. The strategy suggested in this chapter raises many questions about existing education delivery. Many of the answers involve decisions about cost and payment and the job description of educators. It should be noted that the suggested strategy is intended to complement existing traditional and online education with the hope of making the experience better. "For the university, the transition to an Internet-based learning environment requires a restatement of institutional missions and priorities, a revision of conventional structures. For the instructor and student, online courses represent a shift in educational philosophy and instructional design as the emphasis moves from 'teaching' to 'learning', leading to a student-centered, rather than instructor-based system. The challenge for higher education is to find the best way to adjust to this paradigm" (Onay, 2002).

The strategy discussed in this chapter proposes that the university provide, sponsor, administer and maintain an automated online portal to post and sell faculty-created material. An "e-store," selling products developed by the university's faculty members. The web site could be organized like the university's bookstore. The university would act as the "publisher" for the faculty. This approach would also maintain the traditional university's territorial position and claim to resources.

The proposed strategy has three potential benefits:

- Improve the quality and expand the amount of resources for university teaching/learning.
- Provide the developer with a financial incentive.
- Provide the service provider with a financial incentive.

The purpose of this chapter is to outline a strategy for faculty development and university online delivery of supplemental university class materials. As part of this strategy, a royalty is paid to the content developer and the university provider. This plan provides financial incentives to both the faculty member and university. This is a potential source of "new money" for the university. It is hoped that universities will explore this idea to develop new ways of packaging and delivering education that better reward the faculty developer, help pay for the service and also add "value" to the education experience.

Potential "Value" of Online Teaching/ Learning Components

Teaching/Learning Components

It has been substantiated by numerous studies that anywhere/anytime access to resources is the greatest advantage (and that the lack of interaction is the greatest disadvantage) of education delivery using the Internet. In order to avoid redundancy to confirm these points, subsequent discussion about "promoting and stimulating" and "assessment and evaluating" learning will build upon these assumptions.

For this discussion, teaching/learning components are considered independent exercises, experiences or activities that are incorporated together to comprise the class content. They can be done independently or in conjunction with other resources. Regardless of the delivery method, classes are usually a combination of teaching/learning components crafted together by the Instructor and delivered over the duration of the class. The following discussion outlines two primary ways that online teaching/learning components can add "value" to educational experiences.

Promoting and Stimulating Learning

Computer games are popular interactive "engagers," using sound, action and an input device. Players typically are not only trying to do something right, but also trying to do it in the least amount of time. This promotes a form of competition, whether with oneself or with others. Interactive automated instructional designs using competition as an engager have not been explored. This is also an excellent teaching technique to promote assimilating information, making a decision and enacting the solution. As stated by Marc Prensky in his book, *Digital Game-Based Learning*, by marrying the engagement of games and entertainment with the content of learning and training, it is possible to fundamentally improve the nature of education and training for students and trainees (Prensky, 2001).

A "game" approach to delivery of content has great potential to engage the user in many ways to promote and stimulate learning. Especially when most of the users are from a generation that has always played commercially available computer games. The interactive "game" approach can be used to help compensate for the missing Instructor or the lack of perceived necessary interaction. Instructional design can use an interactive "front-end" to deliver the content to be learned. This approach can be used not only to communicate information, but also to engage and motivate the user. Exercises can be designed to use multimedia Internet capabilities for audio and animation and the computer mouse or keyboard for input or information manipulation. All of this is "housed" in an automated online platform which gathers immediate feedback regarding the user's input. Expanded performance measurement, such as number of tries or time to complete, can be used as motivators to stimulate learning and develop greater proficiency.

Assessing and Evaluating Learning

Commercial portals are taking advantage of advancements in technology and expanded bandwidth to administer online assessment. Student appraisal, career appraisal, personal appraisal, standardized testing and diagnostic assessment are all offered online. Basic operation steps for these assessments include online input from the user, database manipulation of this input and automatic feedback to the user. Delivery and assessment are database driven and very efficient and economical to administer and document. This operation strategy has the potential to be a very suitable model for online education delivery and assessment. It also has great potential for "real time" training and online assessment of proficiency. Traditional assessment only measures correctness, not time and attempts needed to successfully complete the exercise. A fast time with minimal tries can demonstrate proficiency, as well as subject mastery.

One of the greatest features that can be included in an interactive, automated online experience is immediate feedback to the user. The delivery can be set up so that the user knows immediately if a choice that he/she made is correct. This is an excellent way to use assessment as a teaching mechanism. This is very different from the traditional paper assignment where the user turns in a hard copy and does not get it back graded for perhaps several days. By then the user must refocus on the topic. The immediate feedback that can be built-in to the exercise using automation adds great value to the learning process. The learning is still content-based, but the interactivity offers the user a much more engaging, real-time, feedback-influenced, self-paced learning experience. This approach is appropriate for many subjects, particularly visual-based processes, such as laboratory or simulation exercises.

Examples of Online Teaching/Learning Components

Interactive, Automated Construction Teaching/Learning Components

The following is a brief description of three online automated interactive exercises using immediate feedback as part of the instructional design. A brief explanation follows each. A class assignment to complete these online exercises was made to 34 undergraduate University of Oklahoma Construction Science students in the Fall 2001 Construction Administration class. This group of users was selected because of their common standing in the construction program. Participants had completed the Materials and Methods and Construction Equipment and Methods classes covering information about tilt-up construction and crane parts. It was known that this group of students had minimal exposure to project management strategies and that the Project Management exercise would be the most challenging. This group was also selected because of their known access to workstations with Macromedia Shockwave Player and computers with speakers.

Based on study results, users overwhelmingly thought that these types of exercises were appropriate for construction education. "The unanimous perception of students and Instructors that these types of exercises are appropriate for construction education is a strong reason for further exploration and development. The automated delivery is ideal for incorporation into existing construction class contents and structures. Interactive components can be used to engage the user differently than traditional exercises" (Ryan, 2002). The majority of the users felt that these types of exercises were appropriate for assessment and testing. Most users felt the interactive component enhanced their involvement. Most would rather do online interactive exercises than traditional hard-copy exercises. The concrete tilt up exercise was considered the most enjoyable of the three exercises. For further discussion of this study, see the URL in the References section at the end of the chapter.

Exercise 1: Identify the Crawler Crane Parts

The objective of this exercise is to match the correct name of the crane part on the right to the designated part in the image of the crane on the left.

A picture of the crawler crane is displayed on the left of the screen. Ten parts of the crane are marked with red lines with a green dot on the right end. The dot

Figure 1.

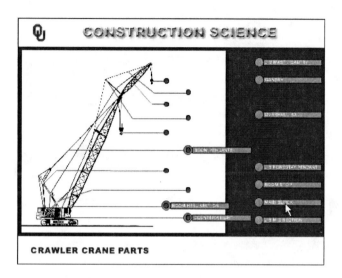

end is a sensitive spot for the drag and drop. Ten blue bars with a dot on the left end are located on the right side of the screen. Each bar is labeled with the name of a crane part. The names are not in order and are randomly located on the screen. This dot end is a sensitive spot for drag and drop. The user is instructed to drag a particular part name and drop it on the appropriate part noted in the image. The user must match the blue dot to the green dot corresponding to the correct part. If the user's choice is correct, the dots stick. If the choice is wrong, the blue bar repels back to its original location. The user can drag and drop the bars in any chosen order. A timer is not visible, but time to complete is shown on the certificate of completion.

This is a content-based exercise paced by the user. It is a matching exercise. The user has to associate all of the right parts to the crane correctly before the exercise is complete. This forces the user to at least read the part name and at least see it on the image, even if for just a moment.

Exercise 2: Sequence the Activities for Concrete Tilt-up Construction

The objective of this exercise is for the user to sequence images of the concrete tilt-up construction process in the order of their occurrence.

Exercises that involve visual recognition for sequencing or arranging activities are ideal teaching tools for construction. This exercise is generic with images easily replaced to depict another activity or set of activities.

Figure 2.

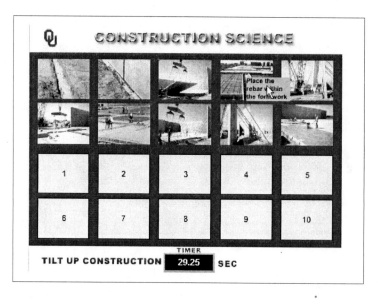

Ten pictures of the tilt-up concrete construction process are displayed on the top of the screen. The user is instructed to drag and drop the pictures into the respective gray boxes numbered one to ten on the bottom of the screen in the chronological order of occurrence. The user must place the images in the correct order of occurrence in the process. If the user's choice is correct, the image sticks. If the choice is wrong, the image returns back to its original location.

"Mouse-over" activated text boxes appear when the mouse is placed on a particular picture. The green text box describes the activity in the picture. The text boxes are hints to help the user differentiate between the pictures and to help sequence them order. The descriptions were added to help users identify details in the pictures due to the limited image size.

A timer is placed at the bottom of the screen to add a competitive element to the experience. The timer activates as soon as the user enters the exercise frame. The purpose of the timer is to motivate the user.

An audio file is activated in the background before the user begins the exercise. Text bullets highlighting the tilt-up construction process transition onto the screen individually as the audio file is played. This is done to refresh the user's thoughts about tilt-up construction before starting the exercise.

As the user performs the exercise every mismatch is recorded. The time and number of tries is posted to the database at the completion of the exercise. The user is given a "Certificate of Completion" at the end of the exercise summarizing his or her performance, including time to complete and number of tries.

Figure 3.

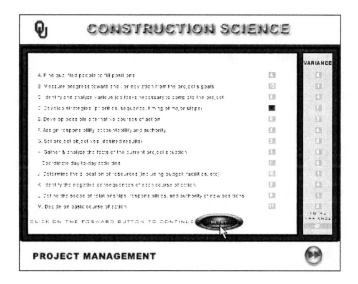

Exercise 3: Project Management Activity Sequencing

The project management exercise is the most complex of the three exercises. The objective of this exercise is to prioritize 13 project management activities based on perceived importance. User priorities are compared to the priority placed on the activity by an "expert."

The user is to assume the perspective of a construction Project Manager. The user types their perceived priority (1-13) in the gray box to the immediate right of the activity. The number one activity is considered to be the most important and is the first that should be done. This exercise is not timed.

The "back-end" part of the exercise is programmed with an expert's views on the priorities that these activities should be assigned. It should be noted that the expert's opinion is a reflection of a particular thought process about the activities. Once all activities have been prioritized and submitted, the "back-end" program calculates the variance of the user's answers from those of the expert.

For example, if the user assigns a priority of five and the expert assigns a priority of eight, the variance is three. The variance is displayed for each activity. The total variance for all 13 activities is displayed at the bottom right of the screen. A perfect score is a variance equaling zero. The user is offered a chance to resubmit a solution. It is hoped that the user will evaluate their initial performance by looking at the variances for each activity and attempt a second try to reduce the total variance.

At the end of the exercise, an audio file explains the expert's logic. The activities transition onto the screen individually as the expert explains why the activity priority was assigned. The user can control the rate of the explanation. The explanation is structured to promote better understanding of the topic. It is hoped that the expert's opinion will provoke thoughts about why the user's approach differed and lead to better understanding of the concepts. The model is an excellent interactive platform for information delivery and thought provocation.

The "Hole" in the Online Component Development Process

Most supplemental online materials and exercises created by faculty are currently offered as part of the content of the "class." Based on the author's exploration of online university class offerings, the majority of these efforts include converted PowerPoint slides, converted document files and links to online resources. Many faculty use class web sites created using university provided "shells" for e-mail and communication. These efforts are very helpful, but do not demand too much time and effort to develop and post. Development of online interactive, automated components like the construction exercises, takes much more time, effort and expertise, but can add great "value" to the education experience. However, currently there is really no focus on the development of teaching/learning experiences like these that utilize emerging multimedia resources and interactive, automated delivery.

As is the case of the development of the construction exercises, the developer absorbed the extra time and effort to develop and coordinate the online resource to try to make the class experience better. It has been recognized for many years that the "hole" in development and use of online class components is that there is a mismatch in the relationship between the time the developer spends creating the resource and the reward for this effort. "By providing incentives such as release time, mini-grants, continuing education stipends, and recognition in the promotion and tenure process, faculty will have more than 'verbal' encouragement to continue, or begin, using distance education technologies and will have reasons to do so" (Murphrey and Dooley, 2000). If the development of these types of experiences is to be promoted, universities should continue exploring strategies to reward developers for this effort.

An observed trend among early university developers of online educational components is to "back away" from further development under the traditional university class model. If faculty are fundamentally paid to "chalk and talk," does the teaching job description require the creation and use of online material? At a minimum, why is using available courseware for communication in a class not mandatory or considered part of the job description? Why is Internet use for teaching not a measured performance category? Based on the author's personal experience, why do faculty have trouble getting credit for this type of effort for tenure and promotion? There is no pay differentiation between faculty that "use" the Internet in their classes and faculty that don't. Because of the questionable support for these efforts, more university developers are seeking other emerging venues to host their work and get paid for the development effort. Commercial portals are being created that cater to the education market, drawing upon the Internet development talents of university faculty. "As the number and type of organizations engaged in Internet-based instruction rises, competition among e-learning providers is likely to increase. Traditional universities are therefore no longer indifferent to the push of new technologies and confront the challenge of redefining their strategies in the 21st century" (Onay, 2002).

The business model of these commercial portals is a royalty payment for each user of the online material, component or class. The evolution of this commercial market is just beginning. If universities do not incorporate a means for faculty to recoup the cost of development effort, these emerging portals, to some unknown degree, will fill that role and reap the benefit. This is worthy of consideration, since most universities already have the infrastructure in place to provide the service. The business part of an e-store can be easily integrated into existing processes.

The University E-Store

The Idea

The university e-store idea evolved after the study exploring perceptions of users about the delivery format, operation, content and potential for use of the three construction exercises discussed previously. Requests by others to use the exercises and the favorable user response led to the idea of selling the exercises as class components. The desire for compensation due to the large amount of effort that was required to create the exercises was a primary catalyst. The whole process of component development, selling and purchase is similar to publishing and buying a book. Components, such as the construction exercises, could be sold in an online store hosted by the university. This idea currently has elevated relevance due to the unfavorable financial status of most universities.

E-Store Products

The items or activities listed below are possible components for a class, another university's class or for an independent user. University faculty members provide many of these components today. All are very suitable for Internet delivery. A component might be required as part of the class syllabus, offered as a supplemental resource or posted for the user's convenience. Bear in mind that not only does the user purchase the component, but also the anytime/anywhere access. Many users would pay a small fee for this convenience alone. The following products might be sold in the e-store:

- web sites
- multimedia exercises
- multimedia labs
- publications
- slide or picture libraries
- course packets
- note packets
- exam reviews
- practice exams
- certification exam reviews
- lecture videos
- guest lecture or special presentation videos

The Business Model

The user goes to the university online e-store portal. The e-store is organized like the university bookstore. Components are accessible by the class in which they are required or organized for purchase by others. Purchase is a typical online transaction using a credit card or charge to the purchaser's university account.

The business model would be the same as a bookstore. The university provides the business end for the Internet portal. Purchase, payment, use and completion of components are automated, requiring no human intervention. The service could be easily integrated into existing university accounting, technology and Internet management processes. For most components the fee would be small, two dollars to five dollars. The user would also absorb the cost of printing or reproduction. As proven by the proliferation of commercial Internet portals built for online purchasing, this business model is very efficient and cost effective.

From the university's perspective new revenue can be generated without a great increase in cost. Existing staff and the in-place online university infrastructure can be used, for the most part. The source of the transaction, the e-store, would be the only addition to the accounting infrastructure that most universities already have in place.

From a developer's perspective the most important part of this model is the payment or reward for the effort to develop and post the online materials. A royalty is charged each time a component is used. This would fill the "hole" in the development process. Though the royalties on a "semester-by-semester" basis are not large, the longer term compensation is a much greater incentive for developing "value" by adding class experiences. The following are additional examples of possible online components and the potential financial incentive for the developer and the university.

Example 1: The obvious place to explore this model is in large, general education classes. Three hundred chemistry students in Chemistry 101 are required to do a series of 10 online interactive virtual labs. Each student is charged two dollars for each use of the 10 online labs. Three hundred students times 10 uses each equals 3,000 (300 X 10 = 3,000) uses during the semester. Three thousand uses times two dollars for each use equals six thousand dollars per semester (3,000 X $2 = $6,000). As a royalty, the developer receives $2,400/semester (40%) and the university receives $3,600/semester (approximately 60%).

Example 2: The instructor creates a web site, http://cns.ou.edu/cns2913/, for the CNS2913 Construction Equipment and Methods class. Content is customized to meet the needs of the class, including text, links, images, video and audio. The web site is not a class, but a component of a class. It is a supplemental class resource, like a textbook. However, the web site is used as the primary reference for lectures and class discussion. The Instructor has requests from other faculty at other universities to use the web site in their classes. Until the web site is published as a book or compact disk and sold, the developer receives nothing for this extra effort and enhancement of his/her class.

The following return might be possible. At the start of the spring semester 40 users pay a $50 fee for access to the web site. It is required in the syllabus, just like a text. Forty users times $50 per user equals $2,000 (40 X $50 = $2,000). The developer receives 40% or $800. The university receives 60% or $1,200. Most faculty will take the extra pay, considering they would probably furnish the web site without reward to enhance the class experience.

There is potential for an interesting situation to develop. If the developer posts the web site on a commercial education portal, can he/she require students to buy access as a part of their university class? The obvious alternative to be explored would be for the university to host the web site in the e-store and reap the benefits, along with the developer.

Example 3: This example is similar to Example 2. A hard-copy Construction Administration Study Guide is required for the CNS 3113, Construction Administration class each Fall semester. The Instructor compiled the study guide, because there was not a suitable text for the course. The hard-copy guide contains text discussion, paperwork examples from an existing construction company in Dallas and other supporting documents. The guide is purchased at the bookstore for the copying and handling fee.

A web site contains the Study Guide, links to other construction administration resources and four online automated reviews that must be completed in order to advance through the web site content. A $45 fee is charged to the user for access to the site. The primary audience will vary between 35 and 45 users each fall semester.

Example 4: A review exam is offered online to all graduating seniors in the Construction Science program prior to sitting for the Certified Profession Constructors Associate Level Exam. All of the seniors are required to take the exam. The review exam is automated, with links to explanations and other information. A $15 fee is charged to take the online review exam. The primary audience will vary between 20 and 30 users each semester. There is also great potential for non-university individuals sitting for the exam to use the review.

Most of the components listed in the E-store Online Components section exist today in some form or another. However, online delivery offers new ways to use these resources. Access 24/7 is probably the most obvious reason that this strategy has merit. "Convenience" is the driver. Many users will be willing to pay two dollars to five dollars to access class components at their leisure or for resources that add value to their education experience. This is particularly true if the components are more engaging than sitting in a class watching someone else do it or talk about it. The cost for the component could perhaps be viewed as a service fee or a tax for not attending the traditional class or needing supplemental resources. This added cost might have a double-edged result.

Some users might react too positively and abuse the availability and not attend classes at all. Others might attend class more regularly, in order to avoid spending additional money.

Selling one component is obviously not a dedicated market to which applying resources is worth the effort. The idea becomes more feasible, however, if there is one dedicated person organizing and programming for 100 class components posted in the e-store each semester.

Observations and Conclusions

Recouping the Cost

Many will view the e-store as "passing another cost on to students." It is. This model suggests a broader approach to providing "value" to education and recouping the cost for the developer and provider. It is different, but not so different that it can't fit into the current approach to university education. Computers were added as part of the university service with "new money" generated from passing the cost on to the students. Creation and delivery of interactive, automated online content is a service that has costs. The cost is being passed on to the faculty and university now, but not to the users.

Compensating the Developer

A primary benefit of this strategy is the opportunity for developers to receive compensation for online development efforts. This strategy answers the question of "reward for effort" that has loomed since the idea of using online materials to supplement education began. The university also creates another "revenue generator" that can be easily integrated into existing business structures.

It is hoped that this simple financial reward system will provide incentive for faculty to create better online material and experiences to enhance education. The unanimous perception that the construction exercises are appropriate for construction education is a strong reason for further exploration and development. Automated online delivery is ideal for incorporation into existing classes. Interactive components can be used to engage users differently than traditional exercises. The project management exercise demonstrates an instructional design provoking thought, as well as assessment, to enhance the learning experience.

Will there be a proliferation of inferior quality class components? Issues of quality and content can be dealt with in a traditional manner. As stated in previous work, "Quality expectations for online and lecture classes should be the same for the classes to be considered equal" (Ryan, 2002). Online class components should be judged with the same considerations as similar class components used now. Like a text, if a resource is good, others will use it, too. It should be noted that like any resource used in a class, selection should be made based on the appropriateness for the teaching/learning objective.

Exploring Instructional Design Using "Adaptive Delivery"

The Internet is an ideal medium for using computer adaptive testing. "With computer adaptive tests, the examinee's ability level relative to a norm group can be interactively estimated during the testing process and items can be selected based on the current ability estimate. Examinees can be given the items that maximize the information (within constraints) about their ability levels from the item responses" (Rudner, 1998). Computer adaptive testing is a "stepping stone" for instructional design incorporating "adaptive delivery." Online teaching/ learning experiences can be crafted so that the type, style and difficulty of information delivered automatically adjust to the skills, pace and understanding of the user. This is a "new frontier" for education.

As this instructional design evolves, so will online products using this delivery. Embracing this opportunity offers a chance for the university to lead in this effort and pay itself and the developer, too. It is also a strong possibility that if universities do not embrace this opportunity, commercial online education portals will.

Opportunity to Use Continuing Education Programs for the E-Store

Most university Continuing Education (CE) programs have online class offerings in place. Typically, these programs have web developers and programmers and host their own servers for online delivery. They are also many times self-supported and an independent entity within the university system. The e-store strategy could be easily incorporated into the CE approach, structure and operation. Most processes and procedures are in place for control, operation and tracking of online offerings. CE could manage the product purchase and access. Quality and delivery can be monitored using in-place criteria. Due to the self-supporting nature of most CE programs, their entrepreneurial motivation will probably be stronger due to self-preservation. These factors provide an ideal environment to explore the e-store strategy.

The University's Role

Many things will influence implementation of an e-store strategy. The commercialism of this strategy for one university or multiple universities as a group is unexplored. It is obvious, because of the blossoming Internet medium, how much more can be done to enhance the learning using images, video, sound, interactive automation, simulation, animation and information linking. However, this development and opportunity to add "value" to teaching/learning comes with a price. Shifting the paradigm to address this cost raises several issues that will prompt diverse reactions from users, developers and providers. It is up to universities to study the merits of this idea. This strategy is another possible step in the on-going development of university Internet education delivery and use.

References

Cooper, L. W. (2001, March). A comparison of online and traditional computer applications classes. *Technological Horizons in Education Journal*, 52-58.

Murphrey, T. P. and Dooley, K. E. (2000). Perceived strengths, weaknesses, opportunities, and threats impacting the diffusion of distance education technologies for Colleges of Agriculture in land grant institutions. *Journal of Agricultural Education*, 41(4), 39-50.

Onay, Z. (2002). *Leveraging Distance Education Through the Internet: A Paradigm Shift in Higher Education: The Design and Management of Effective Distance Learning Programs.* Hershey, PA: Idea Group Publishing.

Prensky, M. (2001). *Digital Game Based Learning.* New York: McGraw-Hill.

Rudner, L. M. (1998). *An on-line, interactive, computer adaptive testing tutorial.* Retrieved November 1998 from the World Wide Web: http://ericae.net/scripts/cat.

Ryan, R. (2002a). Instructional Design and Use of Interactive Online Construction Exercises. *Associated Schools of Construction Proceedings of the 38th Annual Conference.* Retrieved April 11-13, 2002 from the World Wide Web: http://asceditor.unl.edu/archives/2002/ryan02.htm.

Ryan, R. (2002b). *Quality Assurance of Online Courses. The Design and Management of Effective Distance Learning Programs.* Hershey, PA: Idea Group Publishing.

Chapter XIII

Return on Investment for Distance Education Offerings: Developing a Cost Effective Model

Evan T. Robinson
Shenandoah University School of Pharmacy, USA

Abstract

The basis for this chapter is to identify ways in which institutions can maximize their return on investment for distance education offerings through the appropriate and timely re-purposing of the online content for different markets. This will be presented based upon a model of the author's design titled, "Transformative Income Generation," which is a combination of content re-purposing based upon an understanding of the various potential markets in an entrepreneurial manner. The model being presented represents one way to maximize return on investment and, while other ways exist, it is the author's intention to stimulate the reader to consider online educational content, albeit distance education materials, from a different perspective.

Introduction

The intent of action is the achievement of something decisive. When war becomes inevitable, those involved seek to achieve a quick and decisive victory. When a business introduces a new product, it does so with the intent of being successful, profitable and capitalizing on the research and development that led to the product's development. The rougher the market, the more instability in the economy, the more businesses need to capitalize on what works and make sure that it can survive the lean times. Well, higher education is a business operating within an economy that is challenging at best.

The recent state of the economy has led to some challenging times in higher education. The turbulence of the stock market, coupled with challenges of student recruitment and increased competition, has some colleges and universities actively seeking ways to increase revenue. According to a recent report by Standard and Poors, the turbulence of the financial market has placed a financial strain on many institutions due to a decrease in the return on endowment revenues (Peloquin-Dodd and Stern, 2002). The problem is considerably more evident for smaller colleges and universities that have a smaller endowment and a leaner budget than those programs with extremely large endowments that can better weather a drop in endowment contributions. For an example of the financial problems facing higher education, one needs to only read the *Chronicle of Higher Education* to see mention of staff layoffs and cuts to faculty and graduate programs. Given the challenges facing higher education, it begs the question as to how institutions can become more financially stable and less susceptible to the economic rollercoaster that is currently the ride most institutions are on.

The challenge comes in the form of reduced budgets due to financial difficulties and the fact that technology costs can be an easy target when the money gets tight. What inevitably happens is that the resources get reallocated and those programs dependent on technology either limp along with what is currently available or are provided with marginal increases that support the current infrastructure, but do not allow for any improvements. This should necessitate the wise use of reusable resources, one of which is the material developed for distance education.

It has been suggested that by 2004, almost 85% of two- and four-year programs will be teaching online. According to the Council on Higher Education Accreditation (2001) of the 5,655 accredited institutions, 1,979 now offer a form of distance-delivered programs (35%). The escalation of distance education offerings reinforces the reality that for those programs that are already online, more competition is on the horizon and these programs should seek to leverage their position to the best of their ability.

The costs associated with the development of online distance education materials necessitates that the materials must be used as many times as possible to maximize the return on investment based upon the costs to produce these materials. With the marketplace becoming more competitive, it is imperative that the costs associated with delivering online distance education be spread out over multiple markets, as well as over time. This multiple market approach brings to light the idea that entrepreneurship may have a place within higher education and that more consideration needs to be given to multiple markets and out-of-the-box thinking.

The basis for this chapter is to identify ways in which institutions can maximize their return on investment for distance education offerings through the appropriate and timely re-purposing of the online content for different markets. This will be presented based upon a model of the author's design titled, "Transformative Income Generation," which is a combination of content re-purposing based upon an understanding of the various potential markets in an entrepreneurial manner (Robinson, 2001). The various steps of the transformative income generation model will be discussed in detail, along with examples of how it can be applied. Within most sections the considerations or recommendations are subdivided into business-to-consumer and business-to-business approaches, since the transformative income generation model utilizes both. The relationship of entrepreneurship will be discussed as it relates to the transformative income generation model, as will some limitations or considerations to the model, as well. Scattered throughout the chapter are "Helpful Hints," which represent ideas to help with the implementation of the model and the maximization of return on investment. The model being presented represents one way to maximize return on investment and, while other ways exist, it is the author's intention to stimulate the reader to consider online educational content, albeit distance education materials, from a different perspective. According to Einstein, "Imagination is more important than knowledge." This represents an excellent perspective for the reader in that out-of-the-box thinking and using one's imagination will create new perspectives on potential markets and help facilitate the maximization of return on investment for distance education offerings.

Background

The cost of developing online content for distance education has long been a question that has drawn attention as programs consider entering the distance education arena. Some of the costs that have been identified include: technology costs (both start-up and long-term), production costs, recurrent costs (software and hardware upgrades), faculty development costs (conferences, workshops, etc.), faculty instruction costs (compensation for content provided), support costs (help desk, server maintenance), and copyright costs (Morgan, 2001; Morgan, 2001). In essence, anything that can affect the cost of instruction should be included in the determination of what it costs to deliver online distance education. These costs can vary depending upon the type of e-commerce model that is applied and the market that is selected.

According to Hartman et al. (2000), e-commerce can be defined as, "a particular type of e-business initiative that is focused around individual business transactions that use the Net as medium of exchange, including business-to-business and business-to-consumer." It can be further differentiated as being, "... commonly associated with buying and selling information, products, and services via the Internet, but it is also used to transfer and share within organizations through intranets to improve decision making and eliminate duplication of effort" (Fingar et al., 2000). Business-to-business (B2B) models involve the sale or exchange between two business entities while business-to-consumer (B2C) transactions are those in which the consumer is the end-user (Taylor and Terhune, 2001). It is the changing teaching methodologies and the turbulent financial times within higher education that have opened up the possibilities of considering how B2B and B2C models can provide increased revenue to institutions that are willing to take calculated risks. The challenge lies in understanding how to maximize the return on investment for online distance education offerings. The first part to that understanding is to have a firm grasp on what it costs to develop online content.

A number of reference and/or worksheets are available to help determine how much it will cost to deliver online content — an example is found in the handbooks developed by *Western Cooperative for Educational Telecommunications* (2001), which can provide a basis for understanding the challenges that can be faced when integrating technology with the provision of education. The bigger question, the one that is more controversial, is whether it is possible to generate revenue and a subsequent profit via online distance education.

The means to maximizing return on investment for distance education offerings is rooted in a process that evaluates the content in hand, how it is being used in its current form and if it is being utilized to its fullest potential and, finally, if it can be utilized in other ways so that the return on the investment to create online content can be spread over a variety of uses. Designing, creating and releasing content for one purpose to one market is cost ineffective unless that institution has command over a large portion of the market. Applying the transformative income generation model is one way to refute the assumption that the only way to increase revenue via online distance education courses is to teaching to larger classes. With the transformative income generation model, it is possible to maximize the return on investment for distance education offerings by analyzing existing or new content and determining the cost effectiveness of re-purposing it to predetermined markets with a high chance for a solid return on investment.

Transformative Income Generation

When one considers the amount of resources necessary to develop and deliver online distance education, it only seems logical that every opportunity be explored to allow for an institution to capitalize on its investment. Another way to consider this is to consider how many different ways the expense of course development and delivery can be used to generate revenue. The author would argue that most institutions of higher education are content with one, maybe two ways to use content, are constrained by the academic system in which the content was developed, or are reluctant to embrace the "knowledge as commodity" concept.

The transformative income generation model, which will be abbreviated as TIG for the rest of the chapter, is a means by which existing or developed content is re-purposed into other markets to make the most efficient use of the content by spreading the developmental costs over the largest possible number of users. It requires avoiding conservative thinking and embracing out-of-the-box thinking by not emulating your competition but, instead, focusing on areas that they have not.

Within the TIG model, courses or content are considered as a "commodity" which can be sold directly to the consumer (i.e., learner) or to another business. In this respect then, the TIG model allows for the provision of content within both business-to-consumer (B2C) and business-to-business (B2B) contexts. It is about capitalizing on the value of e-content to the largest possible audience. In this way, it might be interesting to think about the drivers within higher education as relate to the TIG model. Should a college or university develop online content and commit resources to the distance education program that supports 300 to 500 students, with the hope that the technology initiative will impact the onsite campus? Or, does that same college or university commit resources to the onsite program supporting 3,000 students and hope the technology initiative impacts the provision of distance education to outside learners? The TIG model is about facilitating this process so that the e-content that is available and re-purposed and helps stimulate change and entrepreneurship while generating revenue.

The TIG model of re-purposing can be developed or implemented on the concept of macro-to-micro re-utilization of educational content so that developing larger blocks of online content in a modular fashion (modules, courses, capstones, etc.) can be used in its entirety as the different subunits to provide a variety of educational opportunities that can appeal to consumers or other businesses. Within any number of disciplines the statements, "This class would make a good continuing education program or certificate program" or "I bet someone else would like this class in addition to our students" can always be heard. But those conversations seldom seem to go any further. Applying the TIG model could help transition these discussions into action items by providing a mechanism that makes it possible to turn ideas into outcomes.

Before introducing the first part of the TIG model, the reader should be advised that maximizing the return on investment for distance education offerings can be accomplished in a variety of ways. The TIG model represents one such means, but as you read the chapter you should also consider how each part of the model might be applied at your institution to facilitate increased revenue from developed online materials. Parts of the model like *Market Stratification* could be done without consideration of the rest of the model and, if applied correctly, could help maximize return on investment. The key is to have an open mind to the possibilities and to challenge yourself to think outside of the box.

Step 1: The Internal Audit

The internal audit should be done at least annually to ensure that financial benefit can be gained from every possible opportunity. Once the audit process has been completed, it may be possible to use the same process to review new courses or content as they become available, as well as using it to plan for which materials should be brought online because of their market potential.

The internal audit of the course(s) and other educational content (continuing education, lifelong learning programs, etc.) can be broken into the following categories: content currently taught online via distance education; content currently taught online onsite, but could be modified for distance education and, onsite, limited online or not-online content. The Content Audit Form should be created with two additional columns, "Re-purpose Potential" and "Market Potential," which will be used during the *Audit Analysis* in the next section. An example of the Content Audit Worksheet can be found in Figure 1.

The audit is not the time to decide which course(s) or content would work best within the TIG model, but to first develop a comprehensive list of classes that fit the generic profile of being taught online. Within an institution that offers a degree online, the TIG model allows for the most flexibility because it has the greatest degree of variability in terms of the number of potential markets. This will be discussed in subsequent sections as to how an online degree-granting institution could broker the entire degree to another institution while simultaneously selling the various courses to other programs, students, or in other forms to other consumers. The reason for conducting the *Internal Audit* is to do the groundwork for the analysis of the content and its market potential, which is done within the *Audit Analysis*.

Figure 1.   Content Audit Worksheet

Content Area	Re-purpose Potential	Market Potential
Taught online via distance education		
BAFI 513 – Investment and Portfolio Management		
BUS 511 – International Business		
MGT 525 – Current Issues in Health Care Mgmt.		
Taught online onsite		
MIS 501 – Decision Making Tools		
MIS 523 – Data Base Systems		
MKT 511 – Marketing Theory and Practice		
Limited online or non-online content		
BUS 515 – Business Law		
ECN 513 – Managerial Economics		
MGT 515 – Human Resources Management		

Step 2: Audit Analysis

The next step in the TIG model is to rank order the three categories by creating an audit matrix. First, rank each area with a score from one to five, with one being hardest and five being the easiest to re-purpose. Then rank order all the fives, then fours, and so on. For the last category that is not taught online or to a limited degree, this might just be rank ordering the courses from the most online to the least online courses. This portion of the Audit Analysis will discern what content can be offered online quickly and without the commitment of significant resources versus the content that would require time and resource commitment before it could be re-purposed. Remember that the purpose of the TIG model is to develop multiple revenue streams based upon the re-purposing of existing online content. This does not negate the re-purposing of content that is not available online for distance learning. It only recognizes that its use and potential benefit must be weighed against the work necessary to bring it to market.

The second part of the Audit Analysis is the ranking of each category based upon perceived marketability and its potential to develop as a revenue stream. This is not to replace the formal market analysis that will be discussed later. This is only done to discern what is available and what might be of value directly to the consumer or to another institution or organization. Consider this as the time to look for the breadth of market potential and not for the depth of market understanding — that will come later. An example of a partially completed Content Audit Worksheet can be found in Figure 2.

Helpful hint: Remember that market potential in the TIG model is both business-to-consumer and business-to-business. This is where the "out-of-the-box" thinking begins and it might be a good idea to involve several individuals in this process. Applying the TIG model is not a sprint, but, rather, a long race requiring time, patience and planning.

As the audit analysis begins to flesh out, it may become evident that there is some content being taught that has high market potential, but limited initial re-purpose potential. For example, a business school might have an online Master in Business Administration (MBA) program that it could re-purpose (and should!), but also has an onsite graduate class in Human Resource Management that is taught somewhat online by an acknowledged expert in the field with a national reputation. While the online MBA program would be the easiest to use within the TIG model, consideration should be given to transforming the Human Resource Management course to an online format pending the faculty member's cooperation due to its market potential.

To summarize, the Audit Analysis provides the opportunity to identify easily reusable content that may have market potential, as well as identifying not-so-easy-to-use content that should be considered for online use at a later date. It is important to remember that this is the first and most general review — the in-depth market stratification and market analysis will take place later. It is also important to realize that some of the most valuable content may not be courses, but other online content in the form of continuing education, undergraduate courses, life long learning offerings, or review workshops.

Step 3: Market Stratification

Seldom is the content that is taught online via distance education, or any content for that matter, only relevant to one specific discipline. Applying the TIG model is about identifying how the content being taught is not only applicable within that specific discipline, but how it also may be taught within other disciplines, as well as being used in units other than the original one developed. In either instance, the TIG model has potential to increase revenue based on business-to-consumer (B2C) distribution, as well as business-to-business (B2B) distribution. While both offer opportunities for increased revenue, they both also have several considerations that should be carefully evaluated. These are discussed within the *Implementation Considerations* portion of the TIG model.

Figure 2.   Content Audit Worksheet (5 = High Repurposability or Marketability; 1 = Low Repurposability or Marketability)

Content Area	Re-purpose Potential	Market Potential
Taught online via distance education		
BAFI 513 – Investment and Portfolio Management	5	5
BUS 511 – International Business	5	4
MGT 525 – Current Issues in Health Care Mgmt.	5	2
Taught online onsite		
MIS 501 – Decision Making Tools	3	4
MIS 523 – Data Base Systems	3	5
MKT 511 – Marketing Theory and Practice	2	5
Limited online or not-online content		
BUS 515 – Business Law	1	3
ECN 513 – Managerial Economics	2	4
MGT 515 – Human Resources Management	2	5

Business-to-Consumer

To provide a better understanding of the business-to-consumer aspect of the TIG model, let's look at the following example. Pharmacology is taught within a variety of disciplines, including — but not limited to — pharmacy, medicine, nursing, physician assistant programs and dentistry. While it is acknowledged that each discipline teaches different variations or to different levels of complexity, the same content is still taught in the different disciplines and is an expense that is borne by each providing program. What if it were possible to use a pharmacology course developed for pharmacy students to teach students from another health profession as well?

Consider the pharmacology course discussed above being taught to 75 pharmacy students, but also being made available to 50 medical students as well. Since the course was initially developed for pharmacy students, more than likely the developmental costs have already been covered. This means that the additional revenue generated is only being reduced by the incremental expenses necessary to re-purpose the content and to deliver it to the end-user. This then produces a greater return on investment for the re-purposed content than the first-run content delivered to pharmacy students. Additionally, once the content has been established online and enters into a recycle and refresh phase, the subsequent development costs drop even more, increasing the overall return on investment due to an increased margin from the pharmacy school use. This is one example of how content can be used within the TIG model.

Considering what to do with the content, as previously stated, requires out-of-the-box thinking. Within the B2C stratification, a number of potential market segments exist. Some stratified markets with examples include:

- *Credit course(s) within the same discipline:* Enroll a larger number of students within the same program or enroll students from other programs.

- *Credit course(s) within other disciplines:* Enroll students from other disciplines, as long as the course objectives are consistent for both programs.

- *Continuing education within the same discipline:* If a particular discipline has a continuing education requirement, then modify content and enroll graduates. (Note: in some disciplines continuing education is a loss-leader or a free giveaway product, especially to alumni.)

- *Continuing education within other disciplines:* Take existing content, modify it, and enroll learners from other disciplines. This has the allure of specialty content, but the enrolling program should determine if specific continuing education credentials are needed.

- *Certificate programming within the same discipline:* If the practitioners within a particular discipline can benefit from certificate programs, which are longer than continuing education and require some mastery examination, then modify the content and enroll graduates. This type of offering usually has a higher unit cost than continuing education, but must have some value to the practitioner to induce enrollment.

- *Adult education or lifelong learning offerings:* Provides the opportunity to reach a very broad and diverse audience, which can be both good and bad. This type of offering usually has a low unit cost to the consumer, so it needs large enrollments. Could be niche-marketed to target only certain groups (i.e., senior citizens, new mothers, etc.).

- *Course or content remediation within the same discipline:* Provide the content or course to students at other programs to allow them to remediate a course that is not offered via summer school or at breaks at their institution.

- *Review for certification or licensure examination within the same discipline:* Provide the content to individuals who are required to sit for a certification, recertification, or licensure examination who feel they would like a refresher course before taking the exam. This could be linked with some form of a practice test as well, which would be part of the time to re-purpose the materials.

The generic B2C list above is not exhaustive in that within each discipline there are a variety of niche markets that could be identified by knowledgeable individuals in that discipline. The key is to think creatively and to realize that until proven otherwise, everyone is a potential consumer.

Helpful hint: Sit down with two or three colleagues and brainstorm the different uses for the content within your discipline, as well as other disciplines or subdisciplines that might offer the coursework that your institution has developed for distance learners. Go online and search the course title or an indicator of the content your institution has developed as yet another means of determining who is teaching the same subject matter.

Business-to-Business

The second component of the market stratification mix is the business-to-business opportunities that might be available when considering the re-purposing of content. Instead of offering the content directly to the learner, the program offers the content to another entity who then conveys it to the end-user, thus changing the role of the program and the face of the consumer.

Using the example of an online Master in Business Administration (MBA) program, the B2B application of the TIG model the program would be the brokering the developed product to another college or university for subsequent resale to the learner. In this example, the MBA program could be sold in its entirety or in its relevant parts. In either case, the contractual arrangement can be structured to allow for either a lump-sum purchase of the content, an annual contract, or a revenue-sharing arrangement where each party gets a set percentage of the tuition generated. The merits of the three types of contractual arrangements are discussed further in the *Implementation* section of the TIG model.

Just like market stratification in the business-to-consumer model, the business-to-business model requires out-of-the-box thinking as well. A number of potential market segments exist with some examples of stratified markets including:

- *Entire program to be offered by another college or university:* An entire program of study, degree conferring, is contracted to another institution for provision to the learner and conference of the degree. Sole responsibility is provision of content and assessment.

- *Credit course(s) to be offered by another college or university:* Course(s) or content are contracted to another institution for use as deemed necessary, ranging from required coursework to electives. Courses are used as-is and the sole responsibility is for the provision of the content and assessment.

- *Resource for credit course(s) to another college or university for subsequent course development:* Course(s) or content are contracted to another institution for use in the development of courses ranging from required coursework to electives. This would be a beneficial arrangement for a college or university that is in program start-up or looking to improve specific offerings. In this model, the contract will most likely be a onetime arrangement and not a royalties-driven revenue model.

- *Centralized resource for another college or university:* Course(s) or content are contracted for use as a resource by another institution seeking to bolster its supplemental learning resources. An example of this would be contracting a course to an institution for use within the library by students and even faculty.

- *Corporate education or training for in-house use:* A partnership is developed with an existing corporation for in-house use of the course(s) or content for education or training. An example of this would be the provision of a marketing course to a pharmaceutical drug company to help educate sales representatives.

- *Corporate education or training for distributive learning:* A partnership is developed with an existing corporation for the subsequent resale of course(s) or content for education or training. An example of this would be the provision of a marketing course to one company that used that material to train the sales representatives for several different companies.

- *Adult education or lifelong learning offerings to be offered by another college or university:* Course(s) or content are contracted to another institution for use to provide adult education materials that the contracting institution is unwilling to provide.

- *Review for certification or licensure examination offered by a professional association:* Course(s) or content are contracted to a local or national professional association that subsequently sells a review for individuals who are required to sit for a certification, recertification, or licensure examination.

- *Continuing education or certificate program offered by a professional association:* Course(s) or content are contracted to a local or national professional association that subsequently sells it as continuing education or certificate programs.

The generic list of B2B ideas is not exhaustive in that various disciplines will have access to markets others do not. For instance, depending on the content and purpose, the market for a school of business would differ from a school of education. Again, the key is to think creatively and to realize that until proven otherwise, every entity, albeit business or academic, is a potential consumer.

Helpful hint: Sit down with two or three colleagues, as well as individuals from the corporate arena, to get their perspective on not only new markets, but the types of offerings in which those markets might be interested. It never hurts to go online and conduct another search, but in this case broaden the approach to include more corporations and nonacademic entities.

Step 4: Analyzing the Market

When the brainstorming and entrepreneurial considerations have run their course and several potential markets have been identified, it becomes necessary to see which markets are worth pursuing. This is the time to go beyond the breadth and look at the depth by sizing up the various markets to determine how well the content and markets match up.

Analysis of the market can begin with reviewing the literature and conducting an in-depth online search of the potential markets and the competition. One way to gauge market interest and to assess the competition is to analyze the offerings and strategic alliances of the competition. This can help determine the extent to which the market is saturated and generate ideas for potential alliances that can be capitalized upon. Other ways in which the market can be analyzed will depend upon whether the model is business-to-consumer or business-to-business.

Within the business-to-consumer model, it may be possible to assess the market by evaluating the specific market strata to discern if there is any information about patronage patterns, how much these individuals pay for various types of services, the nature and cost of the competition, the quality of the competition's offering, the depth of the market, any geographically underserved areas, and the level of commitment present within the market (i.e., translating inquiry to action).

A good example of a market analysis of the B2C model is one that we conducted at Shenandoah University School of Pharmacy to evaluate the market for a nontraditional Doctor of Pharmacy program. A survey was mailed to a sample of pharmacists in a four-state area (Virginia, Maryland, Pennsylvania and West Virginia) to identify if there was interest in a nontraditional program, as well as certificate programming. The results indicated that a significant number of respondents were interested in pursuing either advanced certification or a nontraditional Doctor of Pharmacy degree within the next two years. The market was also analyzed by evaluating the marketplace by researching the existing nontraditional Doctor of Pharmacy programs to determine the strengths and weaknesses of each program, as well as how many would be direct online competitors. The marketplace contained a number of nontraditional programs, but only two were online. Most were regional onsite programs and the majority were more traditional than nontraditional. This not only helped determine the feasibility of entering the market, but also allowed us to accentuate the strengths of our program when marketing it.

Market analysis within a B2B model may be a bit different in that the market analysis may focus more on the potential market than on the competition, given that there may not be any competition. Once again, I will draw upon a personal example for this section. After the nontraditional Doctor of Pharmacy program was implemented, the marketplace within pharmacy education was evaluated. It was determined that there were colleges of pharmacy that would have liked to support their alumni with a nontraditional program, but lacked the resources or expertise to do so. Once interest was identified, a sponsored luncheon was hosted at a national pharmacy meeting to which the deans from all 81 schools of pharmacy were invited. Roughly 20 administrators attended the luncheon, which was a presentation and discussion on collaborative opportunities in pharmacy education. The luncheon resulted in several potential collaborative ventures that did not directly result in a contract, but still hold some potential. Subsequently, one collaborative contract was developed to provide a nontraditional program for another school of pharmacy for only their alumni and pharmacists proximal to the program. When the contract is enacted it will generate roughly $250,000 in revenue annually for nothing more than providing the content to another program. In this instance, the market analysis was a combination of a number of factors, such as the number of pharmacists seeking a nontraditional degree, the number of programs seeking to satisfy their alumni constituency, and the loyalty of pharmacy graduates to the original institutions. The motivating factor to initiate a B2B contract was that a direct-to-consumer program has a limit on the number of students that could be supported, but offering the content through another college or university required less resources and had the potential to generate additional revenue. The re-purposing described above is but one example of how the TIG model could be successfully applied.

Step 5: Implementation Considerations

In many instances, the development of the educational content for use can almost be easier than the administration, coordination, and oversight that may be required to implement the TIG model. It would work best if one or two individuals facilitated the process, tracked the various offerings and assisted in the communications with the various institutional entities that will be involved (business office, registrar, admissions, financial aid, library, bookstore, etc.). This could be a portion of an individual's responsibilities and someone within the administration would work best. The amount of time this takes will be directly related to how successful each venture is and their collective contributions. Care should be taken to ensure that the potential benefits from the TIG model outweigh the time and effort needed to repackage, re-market, and redistribute the materials. For example, it is easy to say that many courses being offered within a college or university can be converted into continuing education. As previously mentioned, continuing education within many disciplines is considered a loss-leader, in that there is a minimal charge or it is given away for free. So in this instance, there may be no revenue or limited direct revenue generated by the continuing education offerings. This is not to say that it might not generate indirect revenue from alumni, but that would be harder to measure. Regardless of whether it is business-to-business or business-to-consumer, some part of the institution's support services will be involved.

It is critical to the success of the venture that during the implementation an effort is made to ensure that everyone is on the same page once the re-purposed content is ready for use. Remember that the consumer could be a learner, a company or organization, or another institution of higher education, and in each case the procurement and entry process must be as seamless as possible, with as little confusion as possible. The necessity of the need for seamlessness is because the definition of quality has shifted in the past 10 to 15 years and has decreased as quality has become increasingly linked to service as well as content. The final reason to get everyone on the same page is that the first time you apply the TIG model the process will be new to your institution, but the result may be new to the buyer as well. So fear of the unknown could exist throughout the entire transaction.

Helpful Hint: During *Market Stratification*, set aside time to meet with the ancillary services that will be involved in order to develop an understanding of what to do, whose responsibility it is, and how to best facilitate the transfer of information and communication. Some things can be handled via e-mail, but this is not one of them. Nothing beats a face-to-face meeting to make sure everyone is on the same page and working towards a common goal. The key is to do this far enough in advance to allow everyone to ramp up for the new product rollout.

Business-to-Consumer Considerations

Every entrepreneurial venture has its share of challenges or considerations. When applying the TIG model there are several areas that should be examined before moving forward. The B2C portion of the TIG model is learner-related and in the event that the content in question is a credit course, then it will require the involvement of a variety of ancillary services, such as the registrar, admissions, financial aid, and library. This area should be addressed in light of potential accreditation issues associated with the provision of online distance education. If the content is continuing education or lifelong learning related, then certainly the business office and possibly the admissions office may be involved, as well as any academic units that provide these types of offerings, like a college of lifelong learning or extended education.

In order to maximize your return on investment using the TIG model for the content developed, it becomes necessary to move the potential student past the interest stage to the commitment stage, which is when financial commitment takes place. For example, when re-purposing the online MBA program discussed earlier, it might be more appealing, both financially and psychologically, for a potential student to enroll in courses that cost less or take less time but could be counted as credit towards a degree. Unfortunately, longer offerings can also present a barrier to a potential student before enrollment even occurs. This is an opportunity that is lost in that an exposure on some level could translate into the potential student enrolling in one or more courses or even in a degree program. Remember, the TIG model is about maximizing the return on investment by translating online content into opportunities, driving down unit production costs while increasing revenue. The determination of the types of offerings that are developed should be based upon the results of the *Market Analysis*, which was discussed earlier in the chapter.

Another means of maximizing a return investment may be providing students with multiple points of entry into the program. Using the TIG model is about capitalizing on the potential of a given product, so the biggest change may be in the deployment of the content, as opposed to a larger re-purposing of the content. For example, instead of having a degree program with a once-a-year admission, students could be allowed to enter the program at more times throughout the year to decrease the program start date as a potential barrier to enrollment. How this is structured will be dependent on the type of distance learning offerings. For example, asynchronous distance learning courses can be provided much easier and more frequently than synchronous distance learning courses that require more direct and structured faculty time.

Business-to-Business Considerations

Within the B2B part of the TIG model, there are once again considerations to be discussed before venturing forth. For example, copyright clearance it is no longer a question of fair use, but the knowledgeable sale of educational content containing copyrighted materials. In order to sell the content to another college or university, the appropriate level of copyright release must be ensured. Since the cost to obtain copyright permission can vary, a nominal amount should be placed in the budget to ensure some degree of financial accountability so that copyright expenses do not become an "unplanned expense." Another concern is that of intellectual property and ownership. The university selling the courses to other programs for distribution may need to compensate faculty or ensure that course ownership has been resolved before this occurs. This should be thoroughly researched both within faculty policies and then outside the college or university to see what other alternatives exist.

The purchasing institution should ensure that the faculty are comfortable with the courses purchased and that these courses meet the accreditation standards for that institution, albeit specialized or regional. This can be resolved by providing the purchasing college or university with a copy of the material for review, the recommendation that the materials be reviewed by faculty in that area of study or by the curriculum committee, and that the contracting institution contact any relevant accrediting groups to understand the ramifications of purchasing content for reuse. Finally, the content needs to be stripped of identifiers as much as possible to ensure that the content does not reflect the college or university of origination, but is more representative of the purchasing institution.

As discussed earlier, there are three primary ways to structure the contract regarding the sale of content to another institution. The first contract model, a lump sum payment, would work best when content is sold for onetime use and does not have the potential to be a renewable means of revenue generation. An example of this is when an institution brokers a course to another institution for onetime use to help get a course started. The second contract model, the annually renewable contract, works well when the brokering institution provides content that is used as a resource that has to be updated at least once per year. One example of this is when a college or university purchases a self-study course for statistics from another school and places the course on reserve in the library for students to use as a resource. The final contract model, the royalty arrangement, is where the purchasing institution pays a fixed percent of the revenue to the brokering institution. Whenever it appears that learners, whether they are students in college, continuing education, corporate training or any other arrangement whereby individuals will enroll for a period of time, the royalty arrangement may be the best alternative.

For example, the college or university brokering the content could negotiate a split of 45% of all revenue for the brokering institution and 55% for the contracting institution. The reason the imbalanced split is that the brokering institution is only providing the content while the contracting institution is responsible for the marketing, admissions, registration, and other functions that must occur when a student enters a program of study. The brokering institution should set guidelines to ensure that the revenue is sufficient to generate a profit and can be worded to state something like, "the contracting institution will be awarded 45% of all revenue generated from tuition charged in the delivery of XYZ MBA program. Tuition costs should not fall below $XXX per hour and the percentage will be awarded based upon the highest tuitional cost during a given semester." The reason for only stating the minimal charge is that the brokering institution stands to gain more as the tuition is raised and the contract does not have to be rewritten each time. A list of the different types of offerings and some possible contract arrangement ideas can be found in Table 1.

Entrepreneurial Considerations during Implementation

Being entrepreneurial is about capitalizing on the strength or expertise of an organization and leveraging that to some gain. The gain usually is financial, although other types of gains exist, as well. In this way it is possible to see that the TIG model and entrepreneurship are very much interrelated. It is not possible to implement the TIG model without being entrepreneurial.

The only way to act upon entrepreneurial ideas and to maximize return on investment is to foster a culture of risk willingness, as opposed to risk aversion. This is not an endorsement of recklessness, but, instead, a charge to be aggressive where warranted and to use speed to market as a competitive advantage. The best of intentions does not move the process any faster, but it can cause an organization to miss a unique opportunity. A key to success is to act quickly and decisively in that an open market today may be highly competitive tomorrow.

Some considerations regarding being entrepreneurial relate well to the TIG, such as:

- Take chances when warranted and do not be risk averse

- Know the market before you act

Table 1. Considerations for Contract Arrangements in Various B2B Models

Type of Offering	Royalty Agreement	Renewable Contract	One-time sale
Entire program offered by another college or university	X		
Credit course(s) offered by another college or university	X		
Resource for credit course(s) to another college or university for subsequent course development			X
Resource for another college or university		X	X
Corporate education or training for in-house use		X	X
Corporate education or training for distributive learning	X		
Adult education or life long learning offerings offered by another college or university	X	X	
Review for certification or licensure examination offered by a professional association	X		
Continuing education or certificate program offered by a professional association	X		

- Plan first, act second, but definitely act
- Be prudent, practical, and use common sense
- Measure gains versus predetermined projections
- Utilize existing online resources for another purpose to minimize the work involved
- Think outside of the box and be creative
- Keep a portion of the revenue where the initiative began to generate future entrepreneurial spirit
- Realize that there are nonfinancial gains to entrepreneurial activities (Robinson, 2002)

The best of entrepreneurial intentions will amount to very little if there is no discussion about what has to be done. The plain and simple truth is that work without marketing is like no work at all.

Step 6: Marketing the Program

For some reason academia has been prone to the "if you build it, they will come" rationale of marketing various programs. It is not that no marketing takes place, only that sometimes it is more of an afterthought than a carefully developed strategy. Whether the marketing is for the business-to-consumer or business-to-business component of the TIG model, marketing should be considered as repeated exposures through various media overtime. One exposure seldom, if ever, will result in revenue, but a focused and planned approach to generating interest will.

The main point is that the marketing should be clear, concise, professional, and targeted for the maximum effect at the lowest possible cost. It would be easy to develop an effective marketing plan if an unlimited supply of money were available. The challenge lies in developing an effective marketing plan with limited resources. For example, advertisements to promote a business-to-consumer product should be developed and targeted to a specific audience so anyone seeing it (hearing it) would know what it means and how to proceed. You don't market to pharmacists in Time Magazine, but you would in the Journal of the American Pharmaceutical Association.

A wealth of information exists on marketing regarding academia and specific to e-learning that facilitates the author's recommendation that external resources be sought if onsite expertise is not available. Consider talking with the personnel from admissions and public relations to assist in the development of a marketing strategy. The only recommendations that the author has on this potentially detailed and complicated subject is to keep it simple, to test developed marketing materials on the potential audience before releasing them en masse, to be persistent, and to fight the frustrations that are prone to happen if results are not realized reasonably quickly.

Some insights from personal experience when trying to market a new online program may be instructive. For the first two years of the program, I used the "five and five" rule, which meant that I tried to get five minutes with any group of five or more people who were either potential consumers of my product or possible partners in the future. In addition, depending on the nature of the product, it may be possible to develop a marketing strategy that could be viewed as concentric circles so the institution markets a program locally, regionally, statewide, and finally nationally (or internationally) in an organized manner using various media.

An example of one marketing strategy comes from when the School of Pharmacy was promoting the new nontraditional Doctor of Pharmacy program for Shenandoah University. I introduced continuing education speakers throughout the four-state area and handed out program information at the same time. The University sponsored booths at the various state pharmacy association annual meetings, which were always followed within two months with a full-page advertisement in the state association journal for pharmacists. A key factor to help make all this possible is that local associations charge less per square foot for booth space than national associations and that state journals charge less for advertisements than do national journals. To reinforce the message, state mailing lists were obtained for the pharmacists in the four-state area and mailings were sent out either several months before or after the state meetings, just to keep the program and the University in front of any potential consumer. The exposure was more focused, but was also more intense. Since the program was just getting started, it was helpful to start in a four state market and then go to a national audience, which is now achieved by running advertisements once to twice annually to pharmacists throughout the state. Additionally, brief articles were written and submitted to various state association journals to explain the program or to update readers about the program and its online attributes in order to provide the potential student with yet another way to learn about the program. A nice feature of these types of articles is that they are free. An article also presents a different impression than an advertisement does, because an article in an association's journal seems to have more legitimacy. Finally, and most important of all, our web site has been modified on numerous occasions to ensure that it provides a good look, is easy to navigate, and has up-to-date information regarding all of our programs.

The types of marketing materials that should be developed depend on the type of market being entered. The marketing for a business-to-consumer market is different from a business-to-business market. The former is mass marketing and the latter is more focused and personalized. Be careful not to overdo the development of marketing materials with too much text, but be even more careful not to fall into the trap of not getting the word out.

Helpful Hint: Review the marketing materials of the competition to get an idea of what you like, what is being done within the potential marketplace, and how you can differentiate your offerings. Gathering these materials can be done by a variety of means, including, but not limited to: visiting a web site, collecting mailings or advertisements, or getting placed on a mailing list. Also, seek the expertise available to you on campus and do not attempt to go it alone in terms of developing a marketing plan, marketing materials, or even so far as to request funding to assist in the initial marketing of something new.

Limitations to Transformative Income Generation

Every idea is prone to having limitations. Within the TIG model, administrative support must be present from the top and allow the academic units or programs the opportunity to make the determinations on what will work. It is only practical that the individual programs, whose personnel are the content experts and understand the potential markets, be able to make the decisions about what can and cannot work. It would be possible to conduct the Internal Audit at the university level, but it could have a profoundly negative impact if the faculty, staff, and administration within each program perceive the intent is to generate revenue from their efforts without their knowledge and with no potential benefit. Keeping empowerment at the local level also helps mediate the next potential limitation, which is what happens to the revenue once it is generated?

It is important to remember that a portion of the revenue stays "local" as an incentive to the individuals who do the work. Leaving behind only a token reward will de-motivate the faculty and staff involved. As word gets out, it also will have a negative impact on anyone with anything to offer. This local incentive does not have to be elaborate and would best be served by doing something that benefits those doing the work, as well as the entire academic unit. For example, when the school of business generates revenue selling its content to another university, then the local incentives could be some new hardware for the individuals doing the content re-purposing and then additional funds for research seed money. This is not a lot of money, but spending a few thousand dollars could help establish buy-in and increase the number of opportunities that can be capitalized upon by fostering a culture of innovation and entrepreneurship. If there is the chance that university overhead costs, as well as other administrative expenses, minimize the amount of funds that are available to the initiating academic unit so it only benefits the university and not the program, then it may be best not to be entrepreneurial. Doing the work and seeing nothing for it will be de-motivating, as well as also creating more financial problems in the long run if the funds generated are considered just another part of the revenue stream and no additional resources are provided and no incentives are forthcoming.

It is imperative that everyone involved realize that applying and benefiting from the TIG model will take time. The worst thing that could happen would be to start the process and not give it enough time to work out. It will take time to re-purpose the content, to develop the new markets and to get the process streamlined and efficient. This is not an overnight process. Care should be taken on the front-end to ensure that everyone involved is aware of this. How long the process takes before it starts to produce results will be dependent upon the flexibility of the institution and of the personnel involved.

The increase in return on investment via the TIG model or any other means should be considered very carefully when applying it to a university budget. The funds generated are soft money, and even if they are the result of a contract, no contract is indefinite and any personnel hired as a result of the contract will need to be supported by other means when the contract expires. Care should be taken to minimize the "fixed" expenses that accompany the new revenue stream by out-sourcing work whenever possible, by supplementing the need for additional staff with better equipment to facilitate the process. For example, if the new facet of revenue is providing content, some of which is on CD-ROM, then the budget for the project should include out-sourced CD-ROM replication or funds for the purchase of a CD-ROM replicator and labeler that can meet the demands of the additional contract, as well as help support onsite activities. It should be clearly understood that it takes money to make money and that the TIG model is not a "something-from-nothing" process.

The idea that it takes money to make money is interesting. It should be made clear that not all types of distance education translate equally within the TIG model. Some means of distance education have more personnel costs than others, which would decrease the margin on the product no matter how many times the content was offered. An example of this is the difference in return on investment that could occur when re-purposing synchronous offerings versus asynchronous offerings. For synchronous offerings, even though the development and associated infrastructure costs may be covered or marginalized over time, the personnel costs are present each time the content is re-purposed, which decreases the margin for that particular offering. Now compare this to the re-purposing of asynchronous content, for which the development and associated costs are covered and the more it is used, the better the margin and the better the return on investment. A key consideration for the TIG model is that online offerings that have larger or repetitive costs should be carefully considered before committing to re-purposing the content.

The final limitation to the TIG model is the marketing that takes place once the opportunities have been identified. It is one thing to identify a potential market. It is something very different to translate that into a contract and subsequent revenue. Commitment and support for the marketing of the re-purposed products must be present for success to result. Once again, the foe to success is going to be not allowing enough time to get the results that are being sought. Remember, marketing is repeated exposure through various media over time. It is possible to get a contract prior to development within the business-to-business model, but that might not work every time in that the content may exist in a form requiring some degree of change, making it difficult to finalize a contract. The key to success is to spend the time talking about the products that have or will be developed and allow time for the word to get out.

Conclusion

The TIG model is a means of generating additional revenue by leveraging existing assets, which within higher education is educational content. The ability of an institution to apply the TIG model will be situational. While some institutions might be able to apply it on a large scale, others may find it works best when applied to a small segment of the curriculum. Instead of applying the TIG model to the entire business school, an administration might consider how to apply it to just three courses in marketing or management. There is nothing wrong with taking the three best offerings available, or even just one, and re-purposing that to help the financial situation of an institution.

Also remember that applying parts of the TIG model will work to your benefit, as well — just work within your means and keep the ideas and plans small and manageable. For example, an institution or program that only conducted the *Internal Audit* and *Audit Analysis* and did not go any further could share those results with the faculty to demonstrate what is online and its potential for the purpose of motivating the faculty and channeling them to focus on developing courses online, both onsite and at a distance. While there would be no direct increase in revenue due to content re-purposing, consider the impact this could have on other revenue streams like student tuition. A program that improves the uses of technology may be able to leverage that within its recruitment of onsite or resident students and maintain, or even increase, the applicant pool. Additionally, the results of the *Internal Audit* and *Audit Analysis* could be used to help focus resources on either successful areas or areas needing improvement, depending on the institution's strategic plan. Regardless of whether the entire model is applied or portions are, hopefully the institution can realize some financial gain, whether direct or indirect.

These financially turbulent times in higher education are not going to go away any time soon. Colleges and universities should seek to secure their financial positions as best they can by utilizing resources available to them and combining efficiencies of content delivery with quality educational offerings. To make this re-purposing work, it must be cost effective and institutions should be cognizant of the high cost to provide technology mediated education and develop ways to maximize the return on investment for what is a highly costly undertaking. Taking a chance is only a risk if no forethought is given. With planning, open communication, and a willingness to venture into the unknown, the TIG model can be successfully implemented.

References

Council for Higher Education Accreditation. (2002). *Accreditation and Assuring Quality in Distance Learning*, Washington, D.C.

Fingar, P., Kumar H. and Sharma, T. (2000). *Enterprise E-Commerce*. Tampa, FL: Meghan-Kiffer Press.

Hartman, A., Sifonis, J. and Kador, J. (2000). *Net Ready: Strategies for Success in the E-conomy*. New York: McGraw-Hill.

Morgan, B. M. (2001). Calculating the cost of online courses. *Business Officer*, 25(4), 22-27.

Morgan, B. M. (2001). *Is distance learning worth it? Helping to determine the costs of online courses*. Retrieved from the World Wide Web: http://webpages.marshall.edu/~morgan16/onlinecosts/.

Peloquin-Dodd, M. and Stern, J. (2002). Weak Equity Markets Hurt U.S. Higher Education Endowments. Standards & Poors, Reprint from Ratings Direct, November 26. New York: McGraw-Hill.

Robinson, E. T. (2001, Fall). Maximizing the return on investment for distance education offerings. *Online Journal of Distance Learning Administration, 4*(2). Retrieved from the World Wide Web: http://www.westga.edu/~distance/jmain11.html.

Robinson, E. T. (2002). A new lesson for eLearning programs: "e" is for entrepreneurship. *Syllabus*, 16(4), 24-25.

Taylor, D. and Terhune, A. D. (2001). *Doing E-Business: Strategies for Thriving in an Electronic Marketplace*. New York: John Wiley & Sons.

Western Cooperative for Educational Telecommunications. (2001). *Technology Costing Methodology Handbook – Version 1.0*. Boulder, CO: Western Interstate Commission for Higher Education. Retrieved from the World Wide Web: http://www.wiche.edu/telecom/Projects/tcm/index.htm.

Chapter XIV

Financing Expensive Technologies in an Era of Decreased Funding: Think Big ... Start Small ... and Build Fast

Yair Levy
Florida International University, USA

Michelle M. Ramim
MIS Consultant, USA

Abstract

The great Greek philosopher Aristotle noted that learning is the outcome of teaching and practice. Clearly, learning is not confined to classroom lectures exclusively. In the past several decades, educators explored the possibilities of providing learning experiences to remote students. With the improvements in technology and the growing popularity of Internet use, online learning caught the attention of both corporations and educational institutions. In this chapter, we will discuss the two common approaches higher education institutions pursue when implementing online learning programs and provide the rationale for their success or failure. Following, we will define, propose, and categorize a set of eight key elements of a successful online learning program implementation in an era of decreased funding. The following chapter also contains a case study about the development of a successful, self-funding, online learning program in the college of business administration at a state university in the Southeast US, followed by a summary and discussion.

Introduction

Traditional learning methodology began transforming when elite universities embraced online education in their degree programs (Forelle, 2003). Progress in distance and online education has increased its popularity in the past decade (Levy and Murphy, 2002). Consequently, it is carving a new brand of universities and causing traditional schools to rethink their business model. Furthermore, some elite schools have developed specialized online degree and certificate programs. In doing so, these schools strive to compete within this new learning methodology and create a new source of revenue, especially due to the declining enrollment and funding resulting from the September 11, 2001 terrorist attack (Roueche et al., 2002).

It is a great challenge to implement a self-funding online learning program, where large seed capital is required to finance such expensive technologies. It is even more challenging to do so in an era of decreased funding, when most schools lack for capital in the first place. The approach taken in this chapter will provide institutions with an understanding of key strategies for a successful, self-funding, online learning program.

The success and survival of a self-funding, online program depends heavily on collaborative efforts to drive the planning and the execution of such challenging initiatives. Starting with a conservative ideology, with a few courses and rapidly advancing to a fully developed degree program is imperative for long-term success. Continuous development of new courses will ensure a steady increase in the volume of students over time. Such an increase is fundamental to generating more funding, which, in turn, should be channeled back into the initiative as an essential element for the continuous growth of the program and its ultimate success.

Background

In the past few decades, universities and colleges have faced a growing demand to attract qualified business students. At the same time, however, universities and colleges are faced with the increase demand by local communities and governments to provide more scholarships and financial aid for local resident and citizen students, in spite of the reduction in financial support allocated to the academic institutions. As a result, higher education administrators have been seeking to increase their overall revenues from corporate sponsors and investors by crafting specialized degree and certificate programs. Not surprisingly, universities and colleges have been relying on international students to compensate by admitting a large number of full-fee paying foreign students (Surek, 2000). Since this has become such an important revenue stream, many business schools have gone beyond designing attractive specialized programs for international students and have even collaborated with international universities around the world to create joint programs. The Wall Street Journal reports that in the year 2001 alone, international students accounted for more than $11 billion for tuition and living expenses in the US, most of which came from abroad (Golden, 2002). Other developed countries experience similar economic benefits from international students. For example, Australia was reported to have gained about four to five billion dollars per year from international students attending higher educational institutions there (Surek, 2000). It was also reported that in the academic year of 2000-2001, the Immigration and Naturalization Service (INS) database contained 547,000 foreign students in the US (Davis and Oster, 2002). The same source reported that the percentage of undergraduate international students in US community colleges increased from 25% in 1996 to more than 36% in 2001 (Davis and Oster, 2002).

In the post-September 11, 2001, era, since some of the hijackers came to the US on student visas, new tougher INS regulations were installed to control and evaluate the issuing of student visas. The *New York Times* suggested that these new regulations would dramatically affect the amount of future international students seeking a US education (Schemo, 2002). That impact is a result of the decrease in overall student visas issued by the INS and the sluggish pace of processing new student visa seekers. The same source also reported that some current international students, who were already in the US, were forced to return to their home country and reapply for student visas under the new regulations. Doing so delayed them for months in their foreign countries while waiting for permission to come back to the US (Schemo, 2002). At the same time, universities outside the US are seeking global recognition from US accreditation bodies and competing to attract such international students. In 2000, AACSB, an accreditation association for business schools, changed its name from "American Association of Collegiate Schools of Business (AACSB)" to "Association to the Advance Collegiate Schools of Business (AACSB) International," to symbolize the shift from an American accreditation body to a more international and global-oriented accreditation body. In 2002, 31 institutions, or more than 7%, out of 430 AACSB accredited business schools were international schools.

Twenty percent of the top 50 M.B.A. programs ranked by the *Wall Street Journal* (Alsop, 2002) were international business schools, including five European, two Mexican and two Canadian business schools. The *Financial Times* also ranked 15 non-US, international business schools in their top 50 executive M.B.A. programs in a 2002 survey. With the increase of global recognition and accreditation of international institutions, augmented by the student visa limitations and the fear of terrorism in the US, an increasing number of students are expected to seek higher education in their home countries.

No one should underestimate the impact of international students on the economy in developed countries such as the US, UK, Canada or Australia, to name just a few. With the student visa limitations and the increased global recognition of international educational institutions, the decrease in funding has become a funding predicament in several US colleges and universities. The *New York Times* reported that commuter students from Canada and Mexico studying in nearby US colleges and universities are banned from entering the US due to the new regulations (Hakim, 2002). These students are not qualified for student visas, as most are not attending American schools full-time. At the same time, they are not meeting the criteria to qualify for tourist visas, either. As a result, some US schools are faced with a substantial loss of tuition, reaching a state of financial crisis.

In the past few years Information and Communication Technologies (ICT), such as online learning, grabbed the attention of many higher education administrators (Levy and Murphy, 2002). In the late 1980s Canadian schools invested an enormous amount of time and resources to develop learning programs for a distance delivery. US schools quickly followed them with some top business schools, like Duke and Michigan, implementing online learning programs in the 1990s. As the use of the Internet increased during the second half of the 1990s, many other US universities headed by their business and engineering schools implemented online learning programs. Almost all included one version or another of an M.B.A. program (Davids-Landau, 2000). In 2000, nearly 25 business schools accredited by the AACSB provided M.B.A. programs entirely online (Arbaugh, 2000a).

Today, more than ever, higher education administrators are very much interested in online learning programs. As they face declining student enrollments, an aging student population, and reduced level of federal, state, and local funding, this has resulted in a growing number of institutions looking for new and innovative ways, mainly by the use of ICT, to attract students in remote or distant locations, including international students (Alavi and Leidner, 2001).

There are two major perspectives about the implementation of online learning programs. The first one is a partnership between higher educational institutions and media or Internet development companies. Several institutions assumed this partnership would be as a successful venture. This assumption was mainly due to the rush most educational institutions and administrators were under to increase enrollment and generate additional funding instantly. In the past half decade, some universities that partnered with private media companies have failed in this type of partnership (Hafner, 2002). Some examples include: New York University that closed NYUonline, the University of Maryland that closed its profit-based online arm in October 2001, and Temple University's Virtual Temple, closed in summer 2001 (Hafner, 2002). Perhaps this type of partnership is difficult to maintain as each entity has its own agenda: the media companies want to increase profits, while higher education institutions first and foremost want to improve (or maintain) their competitive edge and deliver education at a time of plunging state and government funding. AACSB criticized such partnerships, maintaining they can jeopardize the quality, integrity, and significance of the education provided (AACSB, 2002b).

The second perspective about the implementation of online learning programs avoids partnership pitfalls by pursuing in-house capabilities when implementing such projects. Some higher educational institutions find it more appropriate to start a small, in-house, pilot program without commitment to outside media or Internet development companies. The major thrust behind such programs is the "think big, start small, and build fast" approach to eventually offer more robust online learning programs without for-profit partners.

In-house implementation of an online learning program can be challenging, particularly with regard to issues of: administration and institutional support, professional development team, implementation plan, budget and funding, quality assessment and assurance, policies and procedures. Nonetheless, many benefits can result from such an avenue. First, the ability to generate new funding sources by attracting international students who do not need student visas and without the added living expense. Second, the ability to attract students remotely without revenue sharing or a major unexpected hike in peripheral program costs. Third, allowing greater academic control and quality assurance of programs without the pressure for specific, corporate-type profits.

The next section of this chapter will concentrate on some solutions to alleviate the funding crises by implementing in-house online learning programs to attract international and remote students to fee-base programs. A roadmap is presented with some key elements for a successful implementation, followed by a case study demonstrating the effectiveness of such a plan for the College of Business Administration in a state university in the southeastern US.

Key Elements for a Successful, Self-Funding, Online Learning Program

Overview

Like any implementation of information systems, not all are successful or sustaining the point of self-funding. In order to achieve a successful stage of a self-funding online learning program, some investment both in time and capital is essential prior to implementation. It is the assumption of the authors that a successful implementation is a ticket for a self-funding program. Furthermore, in order to avoid online learning implementation failures, it is vital to combine all key elements based on successful cases. This section will present the key elements for successful implementation and a roadmap that can help institutions develop self-funding, online learning programs. It includes justifications for the importance of each key piece for the overall success of the project (see Figure 1). As presented above, the pressure to implement such technologies and programs is growing, especially as a result of the homeland security changes made in the post-September 11, 2001 era. The temptation to quickly implement and offer many online learning courses is undoubtedly present. But in the long run, it is far better to invest time in planning, building administrative and institutional support, as well as evaluating current available platforms. Only then is it recommended to initiate a small-scale implementation pilot. After gaining sufficient experience, establishing a good reputation from faculty and students, along with providing high quality support and development, a major push can be made to build and extend the program to offer more courses and form degree programs online.

In the following sections, we will provide a roadmap to a successful implementation of self-funding, online learning programs, including a comprehensive review of the eight key elements of such an endeavor, beginning with a strategic plan and concluding with quality assurance. Each element is discussed in detail followed by a case study demonstrating the process as it was implemented in a state university in the southeastern US.

Strategic Plan

It is a vital step in any project to take adequate time for proper planning. Without proper planning and adequate time invested in the development of a roadmap, a costly failure can emerge. A good strategic plan should encompass an analysis of all the elements presented in this chapter, along with the development of a

Figure 1. Eight Key Elements for a Self-Funding Online Learning Program

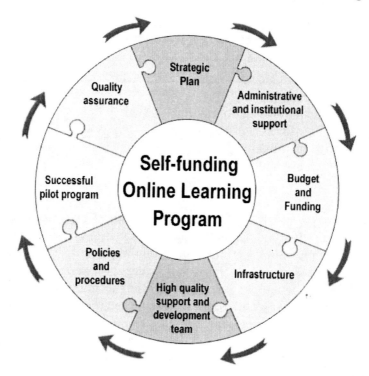

blueprint, or a strategic plan, of the implementation process. Such a plan can also include foreseeable problems and some suggested solutions or avenues of finding solutions to such challenges. It is recommended to propose a plan that is built on a gradual development process, rather than implementing a full-fledged program without proper acceptance by the faculty, students and administrators. With this perspective in mind and the reality of the emerging need for online learning programs and degrees, it is recommended to adopt the "think big, start small, and build fast" methodology. This methodology will allow institutions to progress with their project one step at a time, building and improving based on feedback and constructive comments from users, faculty and administrators. At the same time, the development of a strategic plan for such methodology will provide a solid plan for scalability and funding success.

A thorough assessment is needed for the development of a strategic plan. Such assessment may include: potential sources of funding, potential institutional support, potential students, potential employees, potential faculty, infrastructure (including hardware and software solutions), and the desired timeframe of the project.

Administrative and Institutional Support

Administrative and institutional support is a crucial piece of a successful implementation in the pursuit of a self-funding program. It is important to emphasize to administrators the potential benefits associated with such programs and, at the same time, sincerely present the challenges foreseen in such projects. The lack of education on the benefits and limitation of such technologies may cause some misconceptions about such projects, resulting in roadblocks for a successful implementation.

A major impediment of any online learning program is the lack of initial support from the institution's administration, stemming mainly from misconceptions. It is crucial that department chairpersons, college deans, university provosts or chancellors, as well as university presidents, fully understand the key benefits and challenges of such programs for their competitive advantage.

Many institutions that seek to provide online learning programs invest tremendous efforts in the development of online learning courses, faculty training, and equipment, but lack the overall united institutional support for online learning students. Institutional support for online learning students constitutes all the benefits students enjoy when they are on-campus, but in a format that is available via the Internet or the Web, beyond the access to their online learning courses and interactions with the professor. Such support activities or benefits include online access to: registration, financial aid, the library, the bookstore, advisors, student organizations and virtual communities. In many institutions, most of these support functions are already available and need integration to a single portal enabling students to have access to a centralized point of entry. Such a centralized point of entry, or portal, can be created via the online learning platform where links to the institutional support functions are provided, along with links to the online courses. In the event that some of the institutional functions are not available, it is recommended to investigate and define how such functions can be developed, if possible, during the pilot testing period. As most online learning initiatives begin by attracting mostly current or local students, it is possible to start the pilot program without some of these functions, but maintain a plan to add such features in the near future. In the event an institution is seeking to attract remote, out-of-state, or out-of-country students, it is highly recommended to have all these features linked in place prior to the initiation of the full-fledged program.

Budget and Funding

Funding and budgetary issues are the key challenges for many online learning programs. On one hand, the development of a highly competitive, professional online learning program is dependent on adequate funding for cutting-edge hardware and software, a highly trained staff, and incidental expenses such as: toll-free telephone access for students, and adequate office space. On the other hand, many online learning programs are faced with the challenge of either minimal funding or even insufficient funding to begin with. Such a state can trigger a "domino" effect, where the limited or decreased funding reduces the quality of support, development, and other needed resources (Alavi et al., 1997). Consequently, funding sources must be incorporated as a key component of the implementation plan or be added to current programs to avoid such derailing. One solution for that includes an application for an institution, state, or federal grant to provide the seed funding for the project. Once the project has started and courses are being offered, the project can rely on its own funding and even deliver back some of the funding to the institution.

Online learning courses require a tremendous amount of development, student and faculty support, infrastructure, and other resources, which in most cases are not provided by the university or institution. Therefore, an additional fee is needed to cover such operations. It is recommended to institute a flat fee for all online courses, rather than differentiate and set different fees based on the demand, in the same way universities charge the same tuition fee for business courses as for art and sciences or general education courses. This "across the board" fee for online learning courses should be collected and utilized for the maintenance of the program. It is observed that several universities charge in the range of $250 to $500 over standard tuition for online courses[1]. For private universities, such an additional fee can be added into the general term fees in the same fashion as parking, health, lab fees or "over tuition fees" are added onto students' tuition each term. For state or public universities, the common and recommended way of collecting such funds is by adding an added-value charge for all online learning courses as an "over tuition fee" or computer lab fee. An over tuition fee is defined as an assessed fee charged for certain classes on top of the current state fees to cover the special expenses such a course or program accrued. In some universities, both private and public, some portion of such fee must be shared as their overhead charges, while the rest of the funding can be diverted into the program.

Infrastructure

Software

When looking at in-house development of a customized platform, the school must be aware of both the costs and benefits associated with such avenues. If special features or customized features are needed, a customized solution of some form is probably necessary, although not always the case. We will elaborate further below on a better avenue to pursue. Schools should determine to what degree their desired features are unique or non-standardized and how much effort will be needed to develop such features. If the unique features desired by the school are not addressed by any available commercial packages, a thorough investigation is needed. One aspect of the investigation should look into the costs associated with such development including: time to develop, salaries for developers, upgrades and scalability, proprietary versus open code, training material, etc. The other aspect of the investigation should look at the benefits that will result from the development of a customized platform in comparison to an available commercial solution. If the desired features are standardized, the school is better off with a commercially available platform, rather than going through a system development life cycle.

It is the beauty of Internet technologies and web features that allow integration of tools on a component modeling basis. In other words, when a school, for example, is in need for a unique Web tool to allow students in electrical engineering to perform some virtual instrumentation testing, a module can be developed to accommodate such a feature and be plugged into a commercial package. Hence, it will provide both the benefit of robustness and company support for a steady online learning platform and the benefit of the desired customized feature to accommodate the needs of a particular professor or department. In that regard, it is also possible to combine different modules from different vendors. Although that may cause some technical limitations, the benefits may outweigh the limitations. Institutions can investigate the integration of asynchronous (non-live) commercial online learning platforms, e.g., WebCT™ or BlackBoard™, with synchronous (live) commercial tools, e.g., Centra™, WebEx™, or eRoom™, to provide increased features mostly desired by faculty.

Higher educational institutions are in the business of providing education and not in the business of software development (Levy, 2001). Therefore, a wiser way of pursuing such implementations will be to evaluate the current commercially available packages and determine which platform best addresses the needs of the school, provides better support, allows easy upgrades, integrates best with current platforms on-campus, and is scalable. Choosing a good and credible solution can be a very critical decision. Discussion with users of commercial solutions from other institutions can provide fruitful insights on potential implementation problems or success avenues. Taking the time in examining and evaluating commercial platforms can make the difference between a successful program and a costly failure.

Hardware

Evaluation of existing hardware or a proposal for needed hardware must be included as early as the strategic planning stages of the project. It is common to separate the hardware infrastructure issue into two main avenues: outsourcing and in-house. In the event that competent employees with experience in hardware infrastructure are not likely to be found in-house or may be difficult to find, an outsourcing strategy may be the appropriate solution. It is recommended to investigate the cost associated with large scale operations at the initial stage. Some hosting services will provide a reduced rate for initial implementation, but dramatically increase the rate once the program is growing and expanding both in size of the hardware and bandwidth.

In the event that competent employees can be hired or used in-house, a better avenue can be pursued by the use of existing campus infrastructure (Levy, 2001). Looking at in-house implementation of such a solution requires attention to the subcomponents needed to link the hardware piece. Such subcomponents include: networking and telecommunications, server architecture, along with backup devices and backup policies and procedures. The increase use of network traffic is commonly an overlooked component and may cause a bottleneck effect, thereby reducing the confidence of faculty and students. It is recommended to perform network testing and feasibility analysis prior to implementation to adequately support such a project in-house. The emerging use of minicomputers (or small to medium servers) for e-commence solutions has accelerated the development of server scalability and modular-component-based methodology to hardware support. This allows institutions to start with a small installation and scale up by integrating more servers to the cluster as they grow.

The last hardware subcomponent requiring attention when considering in-house implementation addresses backup devices and policies. It is better to look at backup in the analogy of an "onion." The same way that an onion contains many layers, a good backup plan and procedure will include several layers of a backup process. It is recommended to look at several backup devices and maintain several backup methodologies, such as instant backup to another server, to magnetic tapes, and to DVDs.

High Quality Support and Development Team

Institutions should not spend the time and money on training professors to develop their own courses, just as we don't train vehicle drivers to assemble cars. Rather, they should train professors to effectively use the tools and to deliver their content via the Internet in the same way we train vehicle drivers to drive and obey road signs (Levy, 2001). It is best to assemble a professional support and development team coordinating online course development with the professors and supporting students, rather than have the professors develop their own courses.

In some institutions, a partnership is established with a development and hosting company. One major limitation of such an avenue is that such companies usually do not reveal the pitfalls and complications most likely to emerge during the development and implementation of an online learning program. When time comes, they capitalize on it and increase charges due to the dependence of the educational institution.

A professional support and development team can be utilized as an outside consulting firm on the basis of "per project hire," rather than a partnership as noted above. Another solution can be a team of instructional media developers, programmers, and other support staff employed by the institution. In some cases it might be simpler, and possibly cheaper, to manage a team within the institution, rather than contract an outside vendor. Retaining IT and instructional media developers in educational institutions is often a challenge nowadays. However, institutions can provide some incentives, such as partial or full tuition reimbursement or other benefits in order to retain competent employees. In some cases, using an inside group can provide a great leap, as these employees know the environment and settings of the institution, with a limited learning curve for these issues as compared with an outside group. Using an outside consulting company or Internet course Development Company can be quite expensive and may require considerable coordination efforts. Furthermore, most publishers today provide canned content (electronic package or "e-pack," for short) that is based on the textbooks they have published, thereby extensively reducing the development time of the content.

A good team will eventually include: (a) a program director, the decision maker and motivator of the team; (b) a program coordinator or assistant, to manage the daily operations and serve as a team leader; (c) system administrator(s), one or two team members who are dedicated for hardware, software or infrastructure issues; (d) multiple support staff and developers, to perform the daily support for both students and faculty, along with the development of new content; and (e) graphics and video production team member, to perform the development of professional graphics, audio and video manipulations. If it is also possible, one or two marketing coordinators will help increase the visibility of the program and, in turn, help increase funding.

Clearly, when starting an initiative, it is highly unlikely to have such a robust team. Following the process suggested in this chapter will enable institutions to gradually increase team members and eventually, after few a semesters, reach a point where such a team is in place.

Policies and Procedures

As the first step to establish a high quality program, establishing guidelines for official policies and procedures similarly to the one used for on-campus courses is recommended. Teaching requirements should be defined in terms of academic and professional experience essential for course instruction. Both professors and students should maintain a minimum technical proficiency to prevent the technology from becoming a hurdle in the education process (Levy, 2001). Technical workshops should be developed for newly hired professors, along with a short orientation for students. Best practices show that the use of an online learning system by a faculty member for at least one semester as a web-assisted supplement for their on-campus course can facilitate a smooth foundation whenever the faculty member proceeds to teach online. Some institutions establish a mentoring program where new faculty members are being mentored during their first online term by experienced online faculty. Such mentoring has proven vital in the event other initial trainings or a Web-assisted period is not feasible.

A Policies and Procedure manual usually includes documentation on issues such as:

- Teaching Requirements, including: academic experience, professional experience, technical proficiency, and other competencies.

- New Faculty Hire Mentorship, including: contacts, FAQs, technology setup, best practices.

- New Course Development, including: learning objectives; online assessment types, student motivation techniques, information sources, syllabus development, and textbook selection.

- Facilitating an Online Course, including: course orientations, instructor accessibility, grading, motivational techniques, continuous improvement, and evaluations.

- Student Issues, including: e-registration, e-advising, e-bookstore, e-library, grievance, and online student code of conduct.

Successful Pilot Program

Often, the temptation to launch a full-fledged online learning program is tremendous. Nevertheless, a proper gradual process is recommended, especially in the first two years or so. In most successful projects, such as the example discussed below, following the planning stage is a small scale pilot program. It is recommended to start with Web-assisted or Web-enhanced capabilities for current on-campus courses, rather than with full-fledged online courses. This will allow faculty to learn the system and get comfortable with the technologies without the pressure of swiftly adjusting to a new medium of teaching. It is also recommended to start with faculty who are "first adopters" of technology, as such programs rely heavily on technology and provide close monitoring and support to all students attending online courses.

Web-assisted courses will also allow students to get used to the technology so that when time comes to take a fully-online course, technologies are not a major challenge to overcome, allowing them to concentrate on their learning activities. As time progresses, usually a semester or two following the initial implementation, students and faculty adapt to the system, a higher emphasis should be given to content and pedagogy issues. It is often the case in online learning programs that the "learning" is kept missing whenever the "online" or the advanced technologies takes center stage (Hiltz et al., 2000). By progressing the pilot program with a Web-assisted or Web-enhanced component to on-campus courses, a greater emphasis can be placed on the learning process and engagement, as the technology is not a challenge anymore.

Incorporating feedback and making adjustments to policies and business processes is a valuable outcome of the pilot program. As part of the accreditation process, feedback and adjustment for learning objectives and teaching methodologies (what works and what doesn't) is needed. Doing so following the pilot program not only helps the institution in developing an effective and successful program with the potential for an increase in funds, is also useful for accreditation purposes.

Quality Assurance

Quality assurance involves two components. The first deals with the pedagogy aspect and the effectiveness of the learning experience. The second component deals with the technology aspect and the quality of the product. In the case of an online learning program, the quality of the product will refer to the online course.

Pedagogy

The Association to Advance Collegiate Schools of Business's (AACSB) newly proposed accreditation standards specify that business schools will be required to demonstrate students' achievement of general and management-specific knowledge (AACSB, 2002b). AACSB suggested that business students will meet all accreditation standards, regardless of the delivery mode of the program

(AACSB, 2002b). They also noted that both on-campus and online business programs must strive for the same, or very similar, goals or competencies (AACSB, 2002b). They also suggest that business schools will be required to demonstrate the process by which they judge such learning achievement (i.e., the learning outcomes). Therefore, the pedagogy aspect of quality assurance plays a central role in providing the guidelines for such accreditation requirements.

A high quality course includes clearly defined course objectives and assessment tools, such as the ones used in a traditional on-campus course. However, it is recommended to create incremental assessment as means of encouraging students' learning pace, constant interaction with the professor and discouragement of cheating. Useful assessment measures include weekly quizzes, as well as multiple small and large assignments during the course of the semester. Quality assurance improves over time as the online learning course infrastructure expands and best practices emerge. Continuous improvement and evaluations must be part of the program culture for an increase in effectiveness. As the program progresses, a yearly faculty retreat can facilitate discussion on best practices within the institution and improve the quality of teaching provided by the whole program.

Technology

Technology should not become an obstacle for delivering the educational experience. By the same token that schools are required to maintain their physical classrooms and to make students comfortable when learning on-campus, technology should be transparent and free of errors and down-time periods. The institution should spend time evaluating the current online learning system and seeking one that best matches their expertise and budget.

Once selecting the online learning system, the institution should create a standardized interface, thereby defining the page appearance, institution's logo, the professor's image and biography, and common links to the institution's key sites such as library, registration, and technical support sites. The color palette and banners should also be uniform, as well as a selection of communication tools available to the learners. This will assure consistency and reliability for the learners as they move from one course to another, providing a high quality experience from the technology prospective and simplifying the job of the programming team. This is similar to the uniformly brick-and-mortar classrooms on-campus. The institution can assemble a task force consisting of representatives from the administration, academic instruction and students that will define the appearance of the standard courses. Once the standard course is defined, it is essential that the institution will continue to test and revise the course to ensure high quality and functionality, particularly before the beginning of each semester.

Case Study: Financial Undertaking in an Online Learning Program

In the spring of 1998, the College of Business Administration at a state university in the southeastern US initiated an online learning program. A seed fund of $50,000 was allocated for this initiative with the intent to launch new Web-based courses. At the time, the college was running out of physical space to teach new courses and the idea of teaching over the Internet, while dramatic and revolutionary, was a fitting solution. It complemented the college's aspiration to encourage faculty to blend new technology into the classroom. After evaluating existing tools for online learning implementation, and composing a strategy plan, in July 1998 the college had decided to use WebCT as the platform for the online, web-based course delivery. Effectively, after several meetings with the university computer services department, the college was able to secure a server and some support to run the online learning platform. Thus, at the beginning of the online learning initiative, the support and development team was wholly manned by the university's computer services department. Later, as the project progressed, the college realigned the definition and duties of the support and development team.

The college also spent time working with other university and institutional departments to develop and initiate the program. University Outreach assisted in the establishment of the administrative code for online learning sections in the university system. These section codes are associated with all online courses that enable the "over tuition fee" collection for such special courses. An agreement was maintained, stipulating that the online learning course, which by definition constitutes any off-campus courses, will be registered under University Outreach and will accrue an additional $250 "online learning over tuition fee" for each online course to help fund such an operation. As part of the agreement, the university will keep 10% of the online learning over tuition fee. The rest will be channeled back into the online learning program. To minimize involvement of fee collection, this fee was set by University Outreach in the main university registration system, tacking on the standard tuition once a student is registered for such special sections and collected by the university along with the regular tuition.

By the Fall of the 1998 semester, the first pilot course was launched with 23 enrolled students. The course was taught by one professor and supported by two technical team members and the program administrator. During that term, a progressive development of nine other courses was undertaken and one more technical team member was hired. The college was pleased with the results of the pilot course and the dean was eager to carry on with the project.

By Spring of 1999, the college offered 10 more online courses (fully-online), seven at the undergraduate level, and three at the graduate level. Total enrollment during that Spring of 1999, was 193 students, an average of 19 students per course. Since the college had forecasted low enrollment during typical summer semesters, only five online courses were offered during the Summer of 1999 term, two at the undergraduate level, and three at the graduate level. Total enrollment for the Summer of 1999, was 108 students, an average of 21 students per course.

With the online learning program well under way, the strategy concentrated on extending enrollment growth to 30 students per online course. As a result of the need to support more students, an additional two support-and-development team members were hired. The two included a graphic designer and an instructional designer to assist with the course development. Until that point, most of the support-and-development team members consisted mainly of graduate assistants. These two new members were not students and their salaries were slightly higher. As the goal was to spend time on improving the overall academic quality of the courses, funds generated from the online learning over tuition fees (see section "Budget and Funding" for more information) in the pervious academic year were channeled directly back to fund these activities.

During the academic year 1999-2000, the college began to experience deterioration in the level of support provided by the university computing services department. As the number of courses and students increased, the demand for high quality technical support and system reliability increased. An investigation was underway on ways to shift the online operations from the university servers to local servers at the college. An initial investment of $40,000 was channeled for that project. Appropriately, these funds were channeled from the online learning over tuition fees collected from students taking online courses. As of the Spring of 2000, one of the support-and-development team members assumed the responsibilities of system administrator and the main online learning operations (WebCT, Media server, etc.) for the college shifted to the college level. In spite of the added cost for the operation of the program, the control and ability to provide 24/7 support at a high level was an appealing factor in determining the move. From that point forward, a support-and-development team work schedule was established to manage the help desk between the hours of 8 a.m. to midnight, seven days per week. The transition was smooth, as we continued to work closely with the university computing services. Despite this move, it is important to emphasize that the college continues to rely on the university's computer services department for the overall infrastructure of the networking and telecommunication capacity for the online learning operations, as well as other services, such as e-mail and synchronization of student rosters.

In September 2000, the dean of the college requested to establish a task force to help develop policies and procedures for the program. A small group of faculty who helped initiate the program and the coordinator of the support-and-development team were asked to serve on this task force. The online learning task force has been chaired by the director of the program. Since the academic year of 2000-2001, the group has been meeting twice a semester to discuss general issues related to the program. During the academic year of 2000-2001, the task force was assigned to develop the official Policies and Procedures Manual.

During the academic year of 2000-2001, the college maintained the growth of the online learning initiative and a new Global Executive M.B.A. (GeMBA) online program was initiated. In pursuing ways to provide the best online collaboration to graduate students, a new tool, Centra™, a live-online, learning collaboration tool was acquired. A total of about $60,000 was allocated as a seed fund of the GeMBA program, acquisition of the new tool (Centra), and the development of the first three courses. As in the past, the funds were drawn from the continued operation of the online learning program. Each year the college is graduating a small, but cohesive, group of online Executive M.B.A.s. This year, 14 new GeMBA students will graduate with an accredited M.B.A. degree from Florida International University online. Although seemingly a small group, this cohort includes students from eight different countries, including Argentina, Brazil, Jamaica, Colombia, Mexico, Nicaragua, Venezuela and the US. The GeMBA

Figure 2. Growth of Fully Online Courses in Academic Years 1998-2003

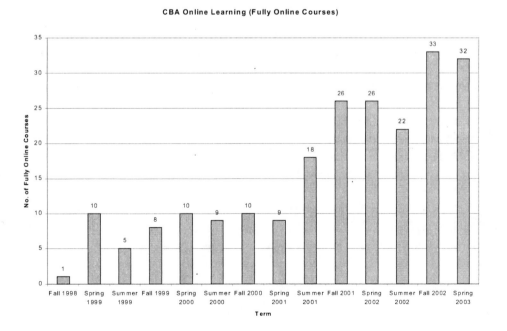

CBA Online Learning (Fully Online Courses)

Figure 3.   Growth of Fully-Online Students in Academic Years 1998-2003

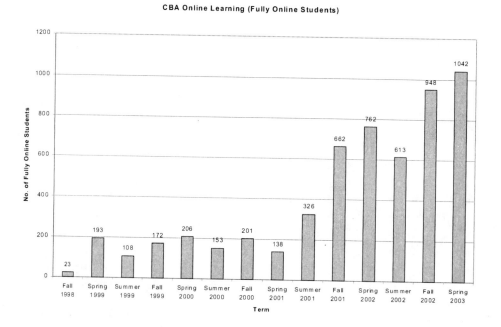

CBA Online Learning (Fully Online Students)

program has become a success story at the university and throughout. A new report that appeared in the December 2000 issue of *Latin Economia* ranked the college's GeMBA program as the 11th best in the US. The rankings, which stem from a study of more than 700 accredited M.B.A. programs, praised the use of technology as the driving force behind this program.

During the academic year of 2000-2001, a new goal was set to develop ten new undergraduate business core courses for online delivery. For this purpose, three additional support-and-development team members were hired and trained. Collaboration with University Outreach on this project was established through their funding of an additional instructional designer who was dedicated to work on this project. By the Fall of 2001, all 10 undergraduate business core courses were completed and two sections of each course were offered, totaling 26 online courses. Total enrollment during that Fall 2001 reached 662 students, an average of 25 students per course. This is evident from the growth in the number of courses from academic year 2000-2001 to academic year 2001-2002 (see Figures 2 and 3). In order to support first-time online students in keeping up with the technology and learning how to use the online learning system, a CD-ROM with videos and tutorials was produced. On average about 300 new CDs are being shipped to students in every term since Fall 2001.

During the academic year of 2001-2002, a new target was set to develop online courses for an undergraduate International Business major. Four more team

Figure 4.   Growth of Web-Assisted Courses in Academic Years 1998-2003

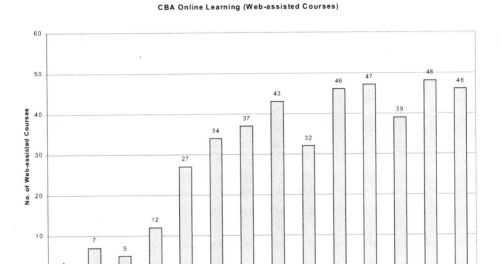

CBA Online Learning (Web-assisted Courses)

members were added to the support-and-development group, totaling ten team members — not including the director of the program. By Fall 2002, four more undergraduate courses were completed and at least one section from each course, along with two sections of the previous ten business core courses, were offered online. At the same time in which the online course offerings expanded, the college, lead by the dean, has continued encouraging all faculty members to utilize the online learning system to supplement their regular on-campus courses (Web-assisted courses). As a result, one full-time team member was assigned to assist faculty with incorporating online learning technologies into their on-campus classes (see Figures 4 and 5). By Fall 2002, the college offered 33 online courses. Total enrollment was 948 students, an average of 29 students per course.

By Spring of 2003, the College of Business Administration operated nearly 80 undergraduate and graduate online and Web-assisted courses combined. More than 1,000 fully online students (not including the GeMBA program) were admitted to the program, generating nearly $700,000 in annual funding for the support of the program, as well as the support of the Web-assisted courses. The online learning support-and-development team provides system administration, development and support for faculty, as well as technical support for students (now consisting of nine full-time and three part-time members) for a total of 370 hours per week. The team includes: a program coordinator, team leader, system administrator, graphics developer, audio and video specialists and other developers. The support-and-development team operates between 8 a.m. to 12 midnight, seven days a week. The support team operates mainly on campus, while some of the support (early morning and late at night) is done remotely. The support requests turnover averages five hours, including weekends. Students are advised to e-mail the support team for any difficulties they encounter along with an available toll-free telephone number. The college continues to collaborate with University Outreach, particularly in matters related to student information services, processing registration and payment collection. University Computing Services provides telecommunication and networking support to the online learning project as the college's vision of virtual learning thrives.

A general overview of the revenues generated from the online learning "over tuition fees," as well as the expenditures associated with running such a program, is available in Figure 6. It is important to note that the seed funding, although calculated in the first term expenditure was a direct investment of the college in such initiative. Furthermore, throughout the first two to three years where the expenditures for supporting the online learning operations surpassed the revenues, the college provided the financial support covering the differences. On the other hand, the online learning program provided some general services back to the college that are not directly related to revenue generating. For example, Web-assisted courses were not charged an "over tuition fee." Yet, professors

Figure 5. Growth of Web-Assisted Students in Academic Years 1998-2003

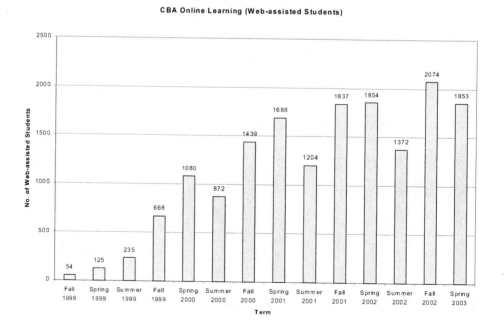

CBA Online Learning (Web-assisted Students)

and students were still able to use the system. It is expected that within two to three years of operations, expenditures will increase at a lower rate than the revenue generation, allowing the online learning program to channel some of the funding back to the college for its initial seed investment, as well as the added financial support throughout the first two to three years.

Summary and Discussion

In this chapter, we provided a roadmap for the successful implementation of a self-funding, online learning program. The tremendous growth and potential of online learning in the business world and corporate training centers in the past few years is making its impact on the academic world. As more and more individuals are stressed for time and demand to expand their education and knowledge, a greater pressure is set on college and university administrators to come up with innovative ways and new technologies to satisfy these needs. With decreasing funding and reduced enrollments, educational institutions are facing a great challenge as such innovative teaching methods and new technologies require a massive seed capital investment, which most schools do not have in the

Figure 6.   Fee Revenues and Expenditures in Academic Years 1998-2003

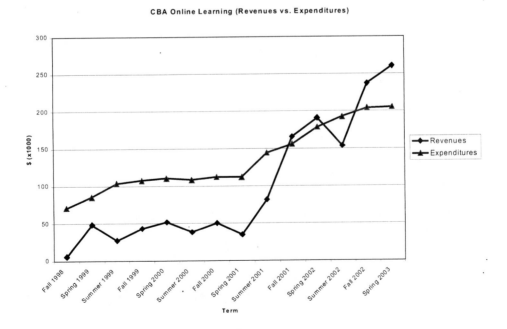

first place. Consequently, higher educational institutions have to spend more time investigating and planning the appropriate methods to implement their self-funding, online learning programs in the era of decreased funding.

Successful implementation of self-funding, online learning programs requires careful research, strategic planning and forecasting. In this chapter a roadmap for success is presented by reviewing eight key elements needed to achieve a self-funding level. The key elements include: strategic plan, administrative and institutional support, budget and funding, infrastructure, high quality support and development team, policies and procedures, a successful pilot program, and quality assurance. The approach recommended by the authors is based on their experience in the development and implementation of several self-funding, online learning programs. A successful case study on the implementation of a self-funding, online learning program is presented, as well as the methodology taken for planning a large-scale, initial pilot program and rapid, steady growth.

References

AACSB. (2002a). *Management Education Task Force Report*. Draft IV.

AACSB. (2002b). *Proposed Eligibility Procedures and Standards for Business Accreditation*. Draft September 12, 2002.

Alavi, M. and Leidner, D. (2001). Research commentary: Technology mediated learning: A call for greater depth and breadth of research. *Information Systems Research*, 12(1), 1-10.

Alsop, R. (2002). The top business schools: Who is up? Who is down? *The Wall Street Journal*, September 8.

Arbaugh, J. B. (2000a). How classroom environment and student engagement affect learning in Internet-based MBA courses. *Business Communication Quarterly*, 63(4), 9-26.

Arbaugh, J. B. (2000b). An exploratory study of the effects of gender on student learning and class participation in an Internet-based MBA course. *Management Learning*, 31(4), 503-519.

Arbaugh, J. B. (2001). How instructor immediacy behaviors affect student satisfaction and learning in Web-based courses. *Business Communication Quarterly*, 64(4), 42-54.

Davids-Landau, M. (2000). Corporate universities crack open their doors. *The Journal of Business Strategy*, 21(3), 18-23.

Davis, A. and Oster, A. (2002). Private eyes bid to help the INS track students. *Wall Street Journal*, February 11.

Forelle, C. (2003). Elite colleges finally embrace online learning. *Wall Street Journal,* January 15.

Golden, D. (2002). Some community colleges fudge the facts to attract foreign students. *Wall Street Journal,* April 2.

Hafner, K. (2002). Lessons learned at dot-com university. *New York Times,* May 2.

Hakim, D. (2002). Immigration policy to bar Canadian and Mexican part-time students in US colleges. *New York Times,* July 9.

Hiltz, S. R., Coppola, N., Rotter, N. and Turoff, M. (2000). Measuring the importance of collaborative learning for the effectiveness of ALN: A multi-measure, multi-method approach. *Journal of Asynchronous Learning Networks,* 4(2). Retrieved May 20, 2003 from the World Wide Web: http://www.aln.org/publications/jaln/v4n2/v4n2_hiltz.asp.

Levy, Y. (2001). E-Learning: An overview of next-generation Internet-based distance learning systems. *Proceedings of the WebNet 2001,* Orlando, Florida, USA.

Levy, Y. and Murphy, K. (2002). Toward a value framework for online learning systems. *Proceeding of the 35th Hawaii International Conference on System Sciences (HICSS-35),* Big Island, Hawaii, USA.

Roueche, J. E., Roueche, S. D. and Johnson, R. A. (2002). At our best: Facing the challenge. *Community College Journal,* 72(5), 10-14.

Schemo, D. J. (2002). Plans on foreign students worry college officials. *New York Times,* April 18.

Surek, B. (2000). Funding problems of technical education in developing countries. (ERIC Document Reproduction Service, No. ED447290).

Swan, K. (2002). Learning effectiveness: What the research tells us. *2002 Annual Report on the Sloan Workshops,* Orlando, Florida, USA.

Valbrun, M. (2002). Bush plan to curb foreigners' studies is scaled back. *Wall Street Journal,* May 9.

Endnotes

[1]   In the case study reviewed in this chapter, the university charged over tuition fee of $250 for undergraduate course and $350 for graduate courses.

Chapter XV

Education Networks: Expected Market- and Cost-Oriented Benefits

Svenja Hagenhoff
Georg-August-University Goettingen, Germany

Michaela Knust
Georg-August-University Goettingen, Germany

Abstract

Discussions about virtual universities, teleteaching or Internet-based learning usually concentrate on pedagogical or technical topics. Ideas and concepts about management, organization or profitability of Internet-based co-operations between universities are missing. This is remarkable due to the ongoing discussion about the efficiency of universities, especially in Germany. This chapter gives an example of an inter-university education network and presents the expected effects of co-operational activities. Both cost-oriented and market-oriented benefits are mentioned. Further on, actual cost- and market-oriented advantages are explained by using the case study of the education network WINFOLine. Finally the authors mention some problems and open questions about how to manage (open) education networks to initiate further discussion about this topic.

Introduction

The use of new media, for example Web-based trainings, is frequently called for by practitioners and politicians in the field of higher education because German universities can only compete on an international level by emphasizing teaching based on information and communication technologies (Hagenhoff, 2003). By using new media in education, teaching is to become more illustrative, motivating, efficient, and, last but not least, less expensive. The idea of saving money, however, should be critically examined because the use of new media is often quite cost-intensive in comparison to conventional instruction at universities, at least if the Web-based trainings are produced at the high quality standards required for self-instruction. To achieve a quality improvement in the field of online education, it is essential to avoid a simple stringing together of pure HTML text pages. A uniform interface of the studying platform, simple serviceability, a good didactic-methodical concept, a meaningful visualization (in the form of short animations), a fullness of exercises and different ways of synchronous and asynchronous communications are just some aspects to be considered during the production of high-quality Web-based trainings. The high costs for the development and operation of such high-quality Web-based trainings often turn out to be a problem, because they usually cannot be borne by a single institution. Besides, it is necessary to reach a large number of students since otherwise the expensive teaching material will not pay off. For this reason it is useful to establish co-operations between education service providers in the field of higher education.

We describe the benefits that arise from such co-operations from the point of view of the co-operational partners. We do not consider potential benefits for the end customers (the students) in this article. The aim of this article is to show the benefits that arise from a co-operation between education service providers in detail. In order to do this, we will first describe the basic types and goals of education networks before turning to the practical example of WINFO*Line*, which will then illustrate the different groups of advantages. Finally, we discuss some open questions, which will have to be answered in the future.

Basics

In the following, we will first give a precise definition of the term "education network" in order to be able to analyze possible advantages. In the next step, we will introduce WINFO*Line* as a practical example.

Ideas and Aims of Co-Operations

The construct of co-operation (Hess, 1999)[1] functions as a starting point for the term network. For years, scientists have used various ways of trying to define the term co-operation. Most of these attempts have come to the conclusion that a co-operation is a tacit or a contractual partnership of convenience between at least two legally and economically independent partners who, for a specific period of time, pursue a common goal in a job-sharing fashion (Grochla, 1972; Knoblich, 1969; Rotering, 1990). With regard to inter-university co-operations, the partners are either associated with universities or interested in the development of educational offers and their contents, this includes individual departments[2] or teachers, as well as companies not directly related to the university, such as partners from the practical field, sponsors or learning platform suppliers.

Through years of research in the field of co-operation, multiple goals, like increases in efficiency or power, as well as time and cost-saving measures, can be taken from the respective specialized literature (Porter and Fuller, 1989; Beck, 1998; Ebert, 1998; Eisele, 1995; Picot and Jaros-Sturhahn, 2001). Often, the list of individual goals is highly enumerative and sometimes the dividing lines are blurred. In the following, the author M. Ebert is thus to be used as an example of someone who has, in a first step, reduced these multiple goals to one prime goal: "The realization of synergy effects." This means that, through the co-operation of multiple partners, advantages which one partner would not have been able to achieve by himself can be achieved. By taking this step, two main groups of goals can be identified: the cost- and market-oriented goals. The cost-oriented goals are primarily reached by economies of scale, which are realized by distributing the necessary tasks on the respective party which is most competent here. Therefore, a more efficient production process can be expected. Market-oriented-goals, on the other hand, can be primarily reached by strengthening their own market position in relation to other parties involved in the market. In a second step, M. Ebert then uses the three basic phases of the process of the increase in value in order to characterize the aforementioned goals in more detail. Input synergies appear prior to the production process and touch upon the fields of purchasing, financing, Research and Development, as well as securing resources. Process synergies appear as part of the actual production process. After this production process, their improved coordination and output synergies, for instance, can be realized through contacts to the market (Ebert, 1998; Hagenhoff, 2002). The partners of such co-operation can gain advantages by aligning the co-operative tasks with these abstract goals. In the following, we will examine these advantages in relation to the main point of interest, namely the inter-university education co-operations.

Case-Study WINFO*Line*

WINFO*Line* is an Internet-based inter-university co-operation between the institutes of information management of the German universities of Göttingen, Kassel, Leipzig and Saarbrücken. The main aim of this inter-university co-operation was to attempt to achieve a high-quality enlargement of the individual curricula by new Web-based trainings. Due to limited resources, the institutes would not have been able to achieve this in the solo run. The co-operation was established in 1997. It was initially founded by the Bertelsmann-Foundation and the Heinz-Nixdorf-Foundation (Ehrenberg et al., 2001). Since 2001 WINFO*Line* has been sponsored by the German Federal Ministry of Education and Research as part of the promotional program "Neue Medien in der Bildung" (New Media in Education). It became an open education network which can be joined by any other university.

The pursued aim of WINFO*Line* is the creation of a university and federal, state, general education network in the German university landscape. The focus of the education network is the Internet-based education in the field of information systems. The Web-based trainings used by all partners should be produced according to high quality standards. The pursuit of quality, as well as economic thoughts, led to the foundation of the education network.

During the first phase, each of the four core teams produced two Web-based trainings, resulting in a pool of eight online courses which were supposed to be used for the primary university education. The four information management departments integrated these online courses into their already existing curricula according to their needs. By adding other, thematically related, courses to their own curricula, the departments took the chance to broaden and deepen their curricula without giving up on the department's independent research and teaching. The departments which offer the actual online course imported by other universities are also responsible for providing numerous support options for everyone, because they have the necessary competency. The Web-based trainings developed by the four departments of information management are collected in a jointly used pool and the students of all the universities involved can use all of these Web-based trainings, regardless at which of the four universities they are enrolled. The credit points acquired in the courses are accepted by each of the participating universities (Scheer et al., 2002). Since WINFO*Line* is an open education network, other education service providers have the chance to broaden and deepen their curricula by participating in this open education network as well.

During the second phase, the focus was and is still on both the exchange of the eight primary university education courses mentioned above and on the production and marketing of saleable education products and offers.[3] The education network WINFO*Line* developed the online-based "Master of Science in Information Systems" as the first marketable education offer. Students from all over the world have been able to participate in this program of study since the fall semester of 2002/2003. The Master of Science is an interdisciplinary program which can be divided into three major sections: namely, information management, computer science and business administration. Especially students with at least one year of work experience, who chose a different major in their primary studies, are targeted with this further education offer. Within 15 months they are able to acquire another internationally accepted university degree in the IT area. This further education course is not only open to full-time students, but also to those in employment, who will benefit from the online-design and finish this course in their spare time. In order to enforce the online character of the course, the amount of time students actually have to be present at one of the four core universities has been reduced to a minimum. As a result, students must only travel to one of the four core universities for testing or for participating in a two-day project seminar. Therefore, students from Japan, Russia and Poland have already been successfully admitted into the program.[4]

The four institutes of information management of Göttingen, Kassel, Leipzig and Saarbrücken operate the organizational and technical infrastructure for the open education network and the Master of Science in Information Systems as a core team and make sure that the established quality standards (e.g., regarding the didactic-methodical arrangements of the courses or the support scenarios, like mail support, chats, hotline-service, etc.) are met by the new partners (Hagenhoff, 2002).

Inter-University Co-Operations

Types of Inter-University Co-Operations

Through more than five years of experience with the education network WINFO*Line,* as well as through multiple research projects, we have identified two different forms of inter-university co-operations: organizational education networks and networks solely focusing on education. However, mixtures between different forms of co-operations are also possible.[5]

Organizational education networks are characterized by an extensive organizational structure which includes the entire business process of an education provider (e.g., from the input interfaces to the production process up to output interfaces). The organizational education networks appear as one unit on the educational market. Consequently, they have the opportunity to amortize production costs through distribution of the educational products or, rather, the complex educational offers. Figure 1 shows that the business process is based on job-sharing between the individual partners of the education network depending on competence and capacity. Within this network a supervision panel makes strategic decisions,[6] whereas performance managers are responsible for those tasks that are necessary to produce the educational products and to distribute them on the market.

In order to refinance the organizational education network, it can offer both standardized or individual educational products (EP) without opportunities for tutoring or examination, as well as complex target-group specific offers for education, such as complete further educational degrees (Master of Science in Information Systems). While configuring these complex offers for education, the organizational education network has to make make-or-buy decisions to decide

Figure 1.   Organizational Education Network

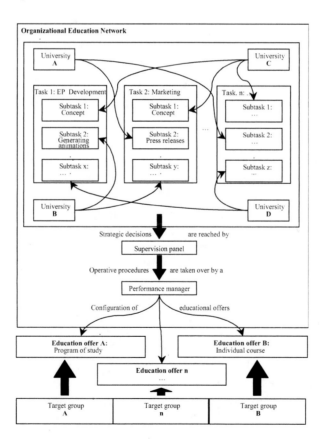

Figure 2. Sharing and Exchanging Material or Courses in an Educational Network

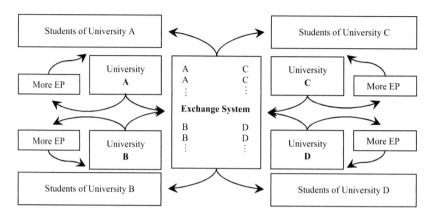

if their own educational products should be completed by additional organizations which are not a member of the network. This can be necessary to furnish the curriculum adequately (Hagenhoff, 2002).

The education network WINFO*Line*, which we have used as a practical example, operates as an organizational education network by clearly distributing the necessary tasks to the individual departments of the four core universities and, thus, designing the marketable Master of Science in Information Systems. The control panel consists of four professors from the different core teams. The research assistants from the four information management departments function as performance managers.

In contrast to the types of networks described above, there also is a form of education network solely focusing on sharing and exchanging class materials between the different network partners. This type of network is illustrated in Figure 2.

In this form of co-operation, the educational products of the participating universities are accessible through an exchange system. The students of this inter-university co-operation have thus the opportunity to not only take classes at their own university, but also to partake in classes offered in the exchange system.[7]

This type of network would also work for the practical example of WINFO*Line*, which would then combine the aforementioned forms of co-operation. On the one hand, WINFO*Line* functions as the aforementioned exchange system between the four departments of the core universities which can be identified as a network solely focusing on education. On the other hand, WINFO*Line* also represents an organizational network in terms of configuration and marketing for educational offers such as the Master of Science in Information Systems.

Expected Benefits of Education Networks

In this section we will describe which benefits can be achieved by an inter-university education network. We will describe the benefits in a generic way. In order to get a structured portrayal of the advantages or benefits within education networks, we will use a simplified value chain, which can be divided into three phases (input, production and output) as a basis.

Market-Oriented Benefits

Market-oriented benefits mainly arise from the stronger position on the market, which can be achieved with respect to suppliers, to buyers and competitors alike.

Input

In organizational education networks, the concentration on competency is supported by the distribution of the necessary tasks to individual co-operation partners and the respective specialization of each partner within a network. On the input side, the organizational education network can achieve market-oriented advantages because one or a small number of partners can launch a focused market analysis (concerning market structure or target group) for the whole network. Through this specialization, the determination of the relevant educational product suppliers and educational products demanded on the market becomes more professional (Eisele, 1995).

We assume that a co-operation consisting of a multitude of equal-minded educational product suppliers has a strong outside appeal due to the reputation of the participating professors (for instance, to other educational product suppliers). An advantageous starting point can thus be expected for the acquisition of new educational products and network partners (Picot et al., 2001). At this point, without anticipating the output-side, we must explain that the configuration of complex education offerings needs a multitude of individual educational products (including the respective tutoring) in order to refinance

Figure 3.   Simple Value Chain (Analogue Böning-Spohr and Hess, 2000)

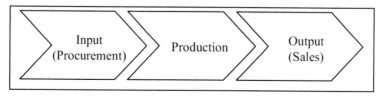

them on the (further) education market (Hagenhoff, 2002). This variety of educational products could be acquired by buying educational offers from other suppliers, for instance. If the percentage of educational products bought elsewhere is extremely high, though, the respective authority (the network) which combines these educational products, transfers them into complex educational products and then sells them gets easily interchangeable. That is why we consider it a better starting point if the education co-operation has a wide range of educational products of their own and then combines them with educational products of other suppliers. If an inter-university co-operation is extended to non-university partners, it can gain additional market-oriented advantages on the input-side, for instance, through a co-operation with the learning platform supplier (Hagenhoff, 2002). Both partners can exchange their know-how, which may result in a win-win-situation. Networks solely focusing on education can gain market-oriented advantages on the input side by making it possible that a multitude of high-quality educational products can be used. Thus, each network partner only has to produce a limited number of educational products. So they can improve their curricula with regard to quality and quantity by the help of other thematically appropriate classes (Hagenhoff, 2002).

Production

Organizational education networks can achieve market-oriented advantages with regard to production primarily by means of time and quality (Picot et al., 2001). The option of specialization through job-sharing within a network supports a faster and more efficient production of educational products. Individual educational products should preferably be designed in a modular fashion in order to be easily used for different types of education (such as further educational courses on the university level or in-house training at privately-owned businesses) with small modifications. The quicker reaction to market needs, meaning the fast realization of an educational product or a complex education offer that is highly demanded on the market (such as an advanced training in the area of business informatics), ensures important competitive advantages (Buchholz, 1996). Additionally, the specialization within the network supports a more professional and more unified quality assurance of all educational products, both the ones they have designed themselves and the ones they have bought from others. The quality standards established by the network can improve its market position if noticed and accepted by the market. Hereby, consumers form certain preferences for these products. A kind of loyalty towards the product is thus developed by the consumers and the prices become less relevant (Meffert, 2000).

In networks solely focusing on education, the actual production take place within the individual departments. Yet, these networks are also able to achieve market-oriented advantages because the educational products are accessible more quickly within the exchange system.

Output

With regard to organizational education networks, one can once again assume that on the output side, co-operation permits the concentration of resources. The respective analyses of possibilities for financing,[8] as well as marketing and public-relations strategies, can be launched more professionally and new markets can, thus, be opened up.[9] Particularly, the access to international markets is improved if the co-operation functions on this level. Through the increased number of educational products within a co-operation (especially with regard to modularly designed educational products), multiple combinations of study contents become possible. These contents can be offered on the market both as simple educational products without tutoring or as complex educational products. The access to potential target markets is, thus, improved (Eisele, 1995). To amortize the production and distribution costs of the educational product and to pass the break even point, a critical mass of consumers must be reached. Certainly, as this critical mass increases, the more valuable (i.e., cost-intensive) the computer- or, rather, Web-based teaching material is. By improving the access to the target market and establishing co-operative market activities, one can assume that this break even point is reached sooner. In addition to that, the aforementioned number and authority of partners participating in the co-operation generates an effect which enforces the co-operation's market position in relation to other competitors and thus raises certain barriers to the market for competing educational product suppliers (Meffert, 2000).

For the networks solely focusing on education, there is another advantage on the market-side closely related to the image[10] and the degree of fame of individual professors and departments: The number of students at all of the participating universities who can be reached with the educational products from one particular university can, thus, be multiplied.

Financial Benefits

The financial benefits from co-operational activities within organizational education networks include the area of advantages of scale, which are created through economies of scale, economies of scope and learning curve. Those education networks can obtain advantages by concentrating individual activities on one or only a few partners connected with the network. They can thus increase the range of this activity or the speed of the respective learning activities (Porter and Fuller, 1989). Additionally, they typically achieve these financial benefits by dividing the necessary tasks on the basis of competencies available.

Input

With regard to the input side, organizational networks can achieve process optimizing in the area of acquisition marketing, for instance, if one partner takes on this task for the whole network, thus avoiding redundant activities. Additionally, those networks can accomplish these financial benefits if costly equipment, which can be used by the whole network, is bought only once (Hagenhoff, 2002).[11]

On the input side, networks solely focusing on education also profit from the cost-saving exchange of high-quality educational products and the respective tutoring. In order to improve the curriculum, both with regard to quality and quantity through additional classes, there is only one alternative: costly teaching assignments which would further strain the universities' budgets.

Production

Within this area of the business process of organizational education networks, one can observe numerous financial benefits connected to research and development, as well as to product design and configuration (Rotering, 1990). With regard to research and development, an education network profits as the number of the participating partners increases. The different perspectives, experiences and competencies enhance the creativity essential for the realization of new educational products (Picot et al., 2001). For these forms of co-operation, we can also assume that benefits will arise if information is shared within networks solely focusing on education. We identified personnel costs as an important cost factor during the production of educational products (Hagenhoff, 2002). The necessary tasks can be roughly divided in collecting and structuring contents, devising a didactic concept, realizing media objects (such as flash animations, graphics, video and audio-cuts, as well as text objects), assembling the educational product and coordinating the whole process (Hagenhoff, 2002). By combining tasks at those locations within the education network where the core competencies are located, substantial learning curve effects — and thus advantages of scale — can be expected. These effects positively influence the financial situation (Hess and Schumann, 2000). Additionally, we can expect more efficiently used capacities, both with regard to personnel and to hardware and software. This is because redundant activities can be avoided and, thus, time and costs can be saved (Rotering, 1990).

Output

On the output-side, education networks can attain financial benefits by multiple uses of individual educational products (Rotering, 1990). Within a co-operation of numerous partners, it is possible to develop a large number of Web- or computer-based contents, which can be, relative to their purpose, combined in different ways in order to make them available on the market for different target groups.

As already mentioned in the beginning, the development of high-quality, Web-based educational offers is rather costly. The number of educational products one supplier can have in store will not be sufficient to configure complex educational offers, such as further educational courses.

Process Comprehensive

Organizational education networks can attain process comprehensive financial benefits by simplifying the adjustment mechanisms of all necessary tasks within a co-operation. We can assume that the adjustments within a co-operation can be realized more easily and, therefore, cheaper compared to non-co-operating organizational units trying to make the same adjustments (Hagenhoff, 2002). The improved access to specialized knowledge and abilities within a network forms another group of advantages (Porter and Fuller, 1989).

Actual Cost- and Market-Oriented Advantages of the Education Network WINFOLine

In the following, we will demonstrate how the advantage categories we have already discussed in theory can be applied to the practical example of WINFO*Line*. These market- and cost-oriented advantages have been achieved by using the exchange system as part of the primary university education, as well as by the introduction of the Master of Science in Information Systems as a further educational degree. Market-oriented advantages, meaning advantages which were realized by strengthening WINFO*Line*'s own market position in relation to others, were less significant than cost-oriented advantages. It turned out to be both sensible and productive, however, to distribute clearly outlined tasks to the respective core team, which already had the necessary know-how.[12] As a result, the development of the curricula, as well as development of the respective outline of studies, the regulations for the conduct of an examination, the admission policy, and the fee scale, could be further advanced at those two departments which had before been able to gain the necessary know-how through different types of restructuring within their own universities (such as the adoption of a department-wide credit point system). Likewise, we were able to allocate the project's technical realization to an additional core team partner who already had the required technical know-how. This procedure enabled us to quickly complete the design of the Master of Science, which could thus be introduced to the market at a point of time when few similar products were available and competitive advantages could thus be achieved. These competitive advantages do not only pertain to the know-how acquired by introducing and carrying out this Web-based further educational course, but also to the pioneer position within the (further) education market, which is connected to that know-how. The brand name "WINFO*Line*" has since been associated with this pioneer position.[13]

Time has proven, however, that it is complicated to distribute tasks which are not clearly distinguishable to the different core teams. The coordination process thus gets too complex and additional coordination efforts eliminate the advantages achieved by the distribution of various tasks. Within the education network WINFO*Line*, each participating partner merely produced two education products as part of their co-operation. This pool of eight education products was then made available for all partners and apart from being used in the exchange system, the education products are also included into the further educational course Master of Science in Information Systems. With regard to this program, these eight education products do not only form the foundation for all course offerings, they also prevent the education network's interchangeability as a mere education service provider. We were also able to take advantage of the outstanding market expertise of professors from all four core departments and have thus encouraged numerous other university teachers to participate in the Master of Science in Information Systems. At this point, there are 15 university teachers from a total of eight universities participating in the Master of Science who have added their own Web- or computer-based courses to the further educational program. Consequently, the students of the Master of Science profit from an excellent range of classes.

With regard to this exchange system, the co-operation has led to an initial cost-oriented advantage: a money-saving exchange of high-quality, Web-based courses was possible because teaching assignments within the exchange system did not require funding. We were soon forced to realize, though, that exchange relations between the different information management departments started to get asymmetrical when they used more services (e.g., support for the WBTs which were part of the exchange system) than they were able to offer. Therefore, we are currently looking into a possibility for establishing an improved exchange system which functions, for example, as a kind of clearing house for the services used and provided (Hagenhoff, 2003).

An additional cost-oriented advantage was gained by buying just one learning platform for all users. To present a uniform learning environment, a studying platform was purchased by the education network WINFO*Line*. This platform can be used for the administration of the courses and for the students of the four network partners in the exchange system, as well as for the newly established further educational course. One individual department would not have been able to afford the acquisition of this complex studying platform on its own and, therefore, would have had to do without important functions. Additionally, cost-oriented benefits can be achieved in the area of technical realization of the new WBT, because the education network WINFOLine shares the assembly tool VLEG[14] for the final assembly of the educational products. This tool was developed by a member of one of the four core teams. In this fashion, substantial "double investments" have been avoided in the area of framework programming.

After all, the available competencies are used jointly within the inter-university co-operation WINFOLine. This is possible due to the establishing of a knowledge intranet which is accessible to every member of the network. Important knowledge can be stored within this intranet, such as information about how to produce flash animations or how to handle the quality assurance. Despite a high turnover of faculty and research assistants, the education network can thus ensure a steady increase of knowledge (Hagenhoff, 2002).

Conclusion and Unsolved Questions

In the previous sections, we portrayed different forms of educational co-operations and the respective goals in a generic fashion. Further on, with the help of a three-step process of the increase in value, we have worked out a multitude of market-oriented and financial benefits. After that we explained some of them with the help of the practical example WINFOLine. The following tables are to give a comprehensive overview of these individual points. Networks solely focusing on education and organizational education networks are not differentiated any further, though.

Yet, in addition to this multitude of advantages, there is also a whole string of open questions and problems concerning inter-university education co-operations which have so far remained unsolved both from a practical and a scientific perspective:

Problems of co-ordination (Porter and Fuller, 1989): The steady necessity to co-ordinate the imminent tasks binds management resources and can possibly result in conflicts of interest. From this perspective, we expect that coordination in polycentric education networks can cause conflicts and high expenditures. Since the opening of the education network WINFOLine in 2001, it is focally managed. In opposition to the polycentric form of co-operation, which was used in the years before, we expected to realize a more effective coordination of activities, as well as a quicker way of finding solutions. Additionally, co-operation-related costs should not be neglected.

Table 1.   Synopsis of the Market-Oriented Co-Operation Advantages

Process level	Input	Production	Output
Market-oriented Advantages	• Focused market analyses (educational product suppliers and demand on the market) • Strong outside appeal • Win-win-situation by co-operating with non-university partners • Broadening and deepening the own curriculum • Using a lot of education products and developing only a few own products	• Time advantages • Quality advantages	• Professional marketing procedures • Respective analyses of possibilities for financing • Opening new markets (even international) • Using a pool of education products to configure different education offers • Attaining break-even-points sooner • Refinancing the activities • Enforcing the market position
Professional outside appeal by concentrating on core competences			

Are the cost-related advantages diminished by time-consuming and costly adjustments within the co-operation?

The inclusion of non-university partners: In what form and to what degree should or must non-university partners be included in the education network at universities in order to adequately cater to the needs of the (further) educational market? One possibility is to co-operate with partners from the field of practical education, which have specially-trained staff at training centers. In this fashion, not only could the scientific side take advantage of the educational products, but also the non-scientific (further) educational side. Yet, to what degree are those partners willing to co-operate with the universities? Because the integration of

Table 2. Synopsis of the Cost-Oriented Co-Operation Advantages

Process level	Input	Production	Output
Cost-oriented Advantages	• Optimization of processes • Acquirement of cost-intense equipment only once for the network • Saving liquidity by exchanging courses and tutoring	• Encouraging creativity in development • Learning curves and economies of scale • Efficient use of capacity by specialization • Saving of time by development of educational products	• Multiple usage of contents
	Avoidance of double investments		
	Avoidance of redundant actions		
	Access and assurance of know-how despite of fluctuation of scientists		
	Simplification of coordination		

non-university partners is quite common in other countries, this problem is especially pressing on the German market for primary and further education.

- Quality standards: At this point, a multitude of quality standards exist in the field of Web-based (further) education. Which standards will prevail in the long run? Or rather, can individual standards either be bundled or brought into line with each other? Who is to ensure a unified quality standard (by standardization or else)?

- Legal claims: If educational products such as Web- or computer-based training are jointly produced within a network, one has to decide who owns the legal rights to these products. Is it the professor who has provided the content or is it the faculty and department who was responsible for visual and audio realization? In this area, there is no unified dispensation of justice. As a result, there is a worldwide necessity to settle this problem (Naquin, 2002).

- Asymmetrical exchange systems (Hagenhoff, 2002; Hagenhoff, 2003): Within networks solely focusing on education which exchange educational products, no money needs to be spent on the usage of these educational products and the respective tutoring. However, if an unequal number of students uses these educational products and the respective tutoring at the individual partner universities, this can result in an unfair distribution of work between the co-operating departments. A model has to be developed which is able realize the symmetric exchange of educational products and the respective tutoring. One could develop a model, for instance, which equates each product with a particular number of points and each department can only receive educational products if it has, through their own offerings, collected enough points.

References

Beck, T. (1998). *Kosteneffiziente Netzwerkkooperation*. Wiesbaden, Germany.

Böning-Spohr, P. and Hess, T. (2000). *Geschäftsmodelle inhalteorientierter Onlineanbieter*. Report no. 1/2000 of the Institute of Information Systems, Dept. 2, Georg-August University Göttingen. Göttingen, Germany.

Buchholz, W. (1996). *Time-to-Market-Management: Zeitorientierte Gestaltung von Produktinnovationsprozessen*. Stuttgart et. al., Germany.

Buzzell, R. and Gale, B. (1989). *Strategien und Unternehmenserfolg*. Wiesbaden, Germany.

Ebert, M. (1998). *Evaluation von Synergien bei Unternehmenszusammenschlüssen*. Chemnitz, Germany.

Ehrenberg, D., Scheer, A. W., Schumann, M. and Winand, U. (2001). Implementierung von interuniversitären Lehr- und Lernkooperationen: Das Beispiel WINFOLine. *Wirtschaftsinformatik*, 43, 5-11.

Eisele, J. (1995). *Erfolgsfaktoren des Joint-Venture-Management*. Wiesbaden, Germany.

Grochla, E. (1972). Die Kooperation von Unternehmen aus organisationstheoretischer Sicht. In E. Boettcher (Ed.), *Theorie und Praxis der Kooperation* (pp. 1-18). Tübingen, Germany.

Hagenhoff, S. (2002). *Universitäre Bildungskooperationen — Gestaltungsvarianten für Geschäftsmodelle*. Wiesbaden, Germany.

Hagenhoff, S. (2003). Tauschhandel an Hochschulen. *wissenschafts-management*, 1, 8-12.

Hess, T. (1999). Unternehmensnetzwerke. *Zeitschrift für Planung,* 10, 225-230.

Hess, T. and Schumann, M. (2000). Koordinator im Netwerk. *iomanagement,* 5, 80-83.

Knoblich, H. (1969). Zwischenbetriebliche Kooperationen. *Zeitschrift für Betriebswirtschaft,* 8, 497-514.

Kraemer, W., Milius, F. and Scheer, A. W. (2000). Eine Dokumentation. In Bertelsmann Stiftung, Heinz-Nixdorf Stiftung (Eds.), *Virtuelles Lehren und Lernen an deutschen Universitäten.* Bielefeld, Germany.

Meffert, H. (2000). *Marketing - Grundlagen marktorientierter Unternehmensführung.* Wiesbaden, Germany.

Naquin, D. (2002, October/November). Online learning: Creating systemic organizational change in higher education. *International Journal on E-Learning,* 33-40.

Oechsler, W. and Reichwald, R. (1997). Managementstrukturen an deutschen Universitäten. *Forschung & Lehre,* 6, 282-285.

Picot, A. and Jaros-Sturhahn, A. (2001). Kooperationen beim E-Learning aufbauen. In A. Hohenstein and K. Wilbers (Eds.), *Handbuch E-Learning.* Köln, Germany. Chapter 3.4, 1-22.

Picot, A., Reichwald, R. and Wigand, R. (2001). *Die grenzenlose Unternehmung - Information, Organisation und Management.* Wiesbaden, Germany.

Porter, M. and Fuller, M. (1989). Koalitionen und globale Strategien. In M. Porter (Ed.), *Globaler Wettbewerb, Strategien der neuen Internationalisierung* (pp. 363-399). Wiesbaden, Germany.

Rotering, C. (1990). *Forschungs- und Entwicklungskooperationen zwischen Unternehmen.* Stuttgart, Germany.

Scheer, A. W., Beinhauer, M. and Gursch, B. (2002). WINFOLine - Vom Projekt zum virtuellen Studiengang. In Bentlage, U. et. al. (Eds.), *E-Learning, Märkte, Geschäftsmodelle, Perspektiven* (pp. 70-93). Gütersloh, Germany.

Schellhase, J. (2001). *Entwicklungsmethoden und Architekturkonzepte für Web-Applikationen: Erstellung und Administration Web-basierter Lernumgebungen.* Wiesbaden, Germany.

Endnotes

[1] This chapter will not examine informal networks which form the basis for the exchange of information and ideas between different suppliers of education products.

[2] The participating organizations are not universities itself but organizations within universities such as institutes or departments. Institutes in Germany are not totally autonomous in the economical and juridical sense but professors are the decision-making authorities concerning steering the own institute (Oechsler and Reichwald, 1997; Hagenhoff, 2002).

[3] The education products can be standardized for anonymous customers or individually developed. We use the term complex educational offers if education products are combined with tutoring and examinations. A complex education offer, for example, is a program of study offered from a university.

[4] Further information is accessible at www.winfoline.de.

[5] Kraemer et. al. (Kraemer et. al., 1997) present a collection of intra- and inter-university co-operations.

[6] Strategic decisions, for example, are admissions of new network partners. Are decisions done by all partners together the co-operation is steered in a polycentric way and each partner has to send an agent into the supervision panel. For decisions done by only one or a few partners the co-operation is steered focally.

[7] Analogous to organizational education networks, this kind of network could be steered focally or in a polycentric way (Hagenhoff, 2002). Important decisions within an educational network are the admission of new partners or quality and design standards for educational products. Decisions about integration of courses into their own curriculum are made by each partner himself.

[8] Such sources for financing can, for example, be marketable education products or offerings or the acquisition of third party funds.

[9] In Germany, universities are not allowed to take money from their students who graduate the first time. Taking fees is only allowed in the case of so called further education. Further education markets can be entered by co-operative configuration of study programs and financial facilities can be developed by this way.

[10] We know the importance of positive image from marketing. A positive image is a basic factor for evaluating the quality of services (Meffert, 2000). A professor's reputation can be affected by images.

[11]   This means for example licenses for hard- and software or equipment for a stand at a fair.

[12]   These tasks include the production of education products, the responsibility for marketing or acquisition activities, as well as the configuration of education offerings.

[13]   Multiple studies have been conducted, particularly in the US, concerning the chances and risks of innovation. The most famous example certainly is the PIMS-study, which has conducted the respective research since 1972 (Buzzell and Gale, 1989).

[14]   VLEG means virtual learning environment generator. It is a tool to assemble Web-based learning arrangements (Schellhase, 2001).

Chapter XVI

Supplemental Web Sites: An Innovative Use of Information Technology for Instructional Delivery

Malini Krishnamurthi
California State University, Fullerton, USA

Abstract

The population of today's learners can be described as being more than 25 years old, with a job and/or family responsibilities. The vast student body requires a flexible program that can accommodate job-related travel, need for a more mobile learning environment and a learning method which may be more entertaining and interactive than the traditional "stand and lecture" method. In the process of innovating with technology in the **college curriculum***, business schools are confronted with the issue of achieving a balance between the issues of "richness" and "reach." While members of the faculty see an opportunity to enrich education, administrators see an* *opportunity to realize economies of scale. Results from this empirical study show that students perceive a face-to-face course supported by a web site to be useful in enhancing their academic performance. Almost all the students made use of the classroom lectures and web site resources and did not feel the need to stay away from lectures.*

Introduction

This chapter is organized around the following topics. The first section presents some figures and facts about the many reasons for pursuing distance education and builds a case for conducting research in the area of Web assisted education. This is followed with a section where information and distance education literatures are reviewed to develop the theoretical foundation for the study. Next, the chapter describes the purpose of this study and then the chapter describes the sample, measures and data analysis techniques used in this study. Following this, the results of the study are presented and a discussion of these results follows. In the concluding sections of the chapter, the strengths and weaknesses of the study are discussed and suggestions for future research are presented.

Statement of the Problem

Business schools are facing rising pressures to increase the value added by their services. To meet this challenge, business schools are undergoing fundamental changes in the way they operate and are continuously seeking ways to create future value (Bailey, Chow and Haddad, 1999). In an effort to determine the forces of change that business schools face, Kemelgor, Johnson and Srinivasan (2000) surveyed deans of business schools. They report that deans from public institutions expect distance learning and using the Internet for research and instruction to be significantly more important in the future. Further, they report that in order to attract good students, educational institutions must continually strive to innovate.

As a step toward innovation, business schools are clamoring to integrate Web-based instruction into the college curriculum, varying anywhere from supplementing face-to-face courses with web sites, to offering a complete M.B.A. curriculum entirely online. A 1999 study reports that even the top business schools in the country have come to offer their M.B.A. programs entirely online (Mangan, 1999). MIT's Sloan School offered their courses online in 1997 (Evans and Hasse, 2001). Corporations encourage continuing education and training and are willing to reimburse tuition for online programs (Evans and Hasse, 2001). Even the World Bank, well known for aiding the development of countries around the world, has embarked on distance education programs (Anonymous, 2001).

This innovative strategy has resulted in an expanded student body that now includes those beyond the traditional college-going student (Schwartz, 1995; Okula 1999), such as those who have completed their education and need to re-skill, those who wish to pursue life long learning (Evans and Fan, 2002) and those who do not have access to a local business school, but wish to pursue higher education without having to leave their job, families and community (Evans and Hasse, 2001; Rosenbaum 2001).

So, given the changing needs of the student and advances in technology, we must be cautious in embracing educational technologies and first fully understand their implications. Maximizing the academic experience of the individual student should be at the forefront of all the initiatives undertaken to incorporate technology in the curriculum. The Campus Computing Project (1999) reports that, "To integrate technology into instruction remains the single most important information technology challenge facing U.S. colleges and universities over the next two to three years." This chapter looks at how Web-based technologies have been incorporated in the college curriculum of a large public institution. Insights are provided by examining the experiences of an instructor in the Information Systems and Decision Sciences department at the California State University, Fullerton.

California State University, Fullerton, (CSUF) is located in Southern California and is in the vicinity of the world famous landmark Disney Land. It is one of the 23 universities that are fully funded by the state of California. There are approximately 28,000 undergraduates and about 1,700 full and part-time faculty. In a study done by two members of the faculty (Reisman and Dear, in the year 2000) about the evolution of Web-based **distance learning** strategies at CSUF, they report that the university undertook the initiative of incorporating distance learning technologies in the Fall of 1997. By Spring 2000 there were more than 600 instructors who had incorporated Web-based technologies in 720 courses, affecting more than 20,000 students. At CSUF nearly 74% of undergraduates and 86% of graduate students are employed and yet 62% of all students take 12 or more hours of course work. So, a typical student at CSUF is a working student eager to receive a quality education at affordable prices from an institution that takes its motto of "Where Learning is Preeminent" seriously.

Instruction can be delivered in several different pedagogical formats. The most personal and labor intensive is a one-to-one instructor-to-student format. In this case, the student is directed through the instructional material personally by the instructor. On the other hand, the least labor intensive and most impersonal is a student self-paced correspondence format, such as the kind one may find from manuals or courses that may be offered entirely online. Somewhere in between these two extremes is a format where the instructor delivers instruction to multiple students in a classroom setting. It is this format that is examined in this study and, more specifically, the researcher examines students in a classroom setting which is augmented with a web site exclusively dedicated for that course.

In trying to make a case for a course dedicated web site, it is worth examining the characteristics of traditional and online courses. In the traditional face-to-face course, students go to campus, sit in a classroom at an appointed hour, talk to the professor and fellow students, and turn in tests and assignments in hard copy format. In the case of a Web-based course, students stay home, log on to a personal computer or a lap top computer, gain access to the Internet, download notes and materials any time, anywhere, stay online for however long they wish to, chat with their professors and classmates, submit test and papers through e-mail or digital drop box.

While the traditional method prevails and will probably continue to do so, students have more opportunities than ever before to go to the Internet at the click of a computer mouse. The NEA (National Education Association) reports that faculty teaching both face-to-face courses and fully online courses found the latter to be better in meeting the educational goals of providing access to information and high quality course material. On the other hand face-to-face courses did a better job of addressing individual learning styles (NEA, 2000). Evans and Hesse (2001) report that 18 to 24 year old young adults found that the online courses expected students to be self starters, demanded too much self discipline and were too flexible, and they deprived them of a campus experience. Rad (2002) reports online instruction unsuitable for courses dependent on experiential learning and laboratory exercises and other interaction-based activities. Ponzurick and colleagues (2000) report that their marketing M.B.A. students preferred online education for reasons of convenience, career and location, but would have ideally preferred face-to-face campus instruction because they perceived this to be more effective and gratifying.

Having a face-to-face course supplemented with a web site can offer a richer learning environment than either one offered alone (McEwen, 2001). So, a Web assisted course which could blend the best in both the approaches of learning, was seen as an ideal way to enrich the learning experience at CSUF and, therefore, it became the subject of study in this chapter. Further, it would give the young adults a good starting point (Rad, 2002; Eastman and Swift, 2001) to get acquainted with the online tools and an opportunity to cultivate appropriate learning behaviors for a fully online course that they might encounter in the future.

Results from this study can have implications for faculty, students, university administrators and designers of educational software. From the point of view of students and faculty, having an experience with a supplementary web site can be the first step in dealing with courses that may be offered entirely online. As the paradigm shifts from being "teacher centered" to "learner centered," students will have an opportunity to identify new learning behaviors that they have to cultivate in order to learn through technology intervention. Members of the faculty will have an opportunity to identify new teaching and assessment strategies that are suitable for teaching with technology. Educational technology designers can have a better understanding of the real-life uses of their inventions. Feedback from students can provide insights for technology design and redesign.

Background Research

An examination of the theoretical and empirical literature in the area of distance education in general and course dedicated web site in particular led to the selection of the dependent variable web site use. Some of the naturally occurring needs of the students were examined in developing factors to predict web site use. These needs are: "the need for a useful web site and a user-friendly software to access the site," "the need to work with fellow students and be independent of time," "the need to manage absences" and "to have an opportunity to do better in tests."

Usefulness of the Web Site and User-Friendliness of the Software to Access the Site

If students believe that the content of a course web site would be of use to them and, further, if they believe that the content can be easily accessed via the software, they are likely to visit the site more often. This assertion is based on the diffusion of innovation theory. In this theory, diffusion means acceptance of an innovation and adoption means a decision to continue full use of an innovation (Everett, 1983, 1986). The theory suggests that beliefs and attitudes toward a technology can be important determinants of technology adoption. The belief that a technology is useful and easy to use can influence the users' attitudes toward that technology and, thereby, their decision to adopt that technology. This belief has been empirically tested and found to be true in the case of computer software (Bagozzi, Davis and Warshaw, 1992), e-mail (Gefen and Straub, 1997), the World Wide Web (Atkinson and Kydd, 1997) and course web sites by Angulo and Bruce (1999) and Arbaugh (2000). Therefore, perceived usefulness of a course web site and the user-friendliness of the software can be expected to lead to increased usage of a web site.

To Be Able to Work Together with Fellow Students and be Independent of Time

Students must be computer and information literate by the time they start their professional careers. Students must not only gain expertise in their chosen field of study, but also be prepared for an environment characterized by online communication and be able to work collaboratively with their colleagues using the Internet and intranet tools.

Companies producing courseware offer a suite of features to support synchronous (same time) or asynchronous (different time) communication among individuals and groups. The features range from very simple systems (e.g., electronic mail) to much more sophisticated systems that facilitate collaborative learning (e.g., chat rooms, discussion boards and virtual classrooms with white boards). Therefore, course delivery via the Internet can facilitate cooperative learning experiences and the development of cooperative learning behavior among students (Karuppan and Karuppan, 1999; Mangan, 1999; Brandon and Hollingshead, 1999). Technologies that facilitate group communication are more than just tools to supplement traditional classroom learning. Students get an opportunity to see the work of other students which can inspire them to emulate their peers. And so a totally different learning environment can be developed which can be very enriching and can lead to effective career preparation (Witmer, 1998).

Because the electronic medium can be flexible and independent of time and distance, many researchers (Arbaugh 2000; Mangan 1999; Ragothaman and Hoadley, 1999; Rad, 2002; Surrency et al., 2001; Lahey, 2002), report that course dedicated web sites can help students to overcome temporal constraints. Students and instructors can benefit from this by scheduling their activities to a time frame that is more convenient to them. Further, students can have round-the-clock access to course materials. DeSanctis and Sheppard (1999) summarize the experiences of students in a one-of-a-kind, global, executive online M.B.A. program. The executive students appreciated the opportunity to learn while they were working, traveling or living at home in a distant part of the world. It was an opportunity for the executives to interact with research faculty and to be associated with university life. Because of the asynchronous nature of the communication medium, it was possible for students to bring in their diverse perspectives to solve business problems and also to be able to share their rich cultural perspectives. The cyber classroom had students from many parts of the world. Therefore, this study posits that the presence of a supplemental web site can bring about collaborative learning and time independence and, thereby, lead to increased number of visits to the web site.

The Desire to be Absent

Having a supplementary web site for face-to-face instruction can be a double-edged sword. While the presence of a rich and well organized web site can invite students, it can also tempt them to stay absent from the classroom. This can result in the unintended consequence of instructors resisting to place course materials on the web site because of the fear of losing students in the classroom. The impact of classroom attendance has received very little empirical evaluation. Karuppan and Karuppan (1999) report little or no impact on attendance when comparing semesters in which the web site was used and the semesters when the web site was not used. This study re-examines this issue of web site use and classroom attendance. The presence of a web site can tempt students to be absent more frequently and this can lead to increased visits to the web site.

Opportunity to Improve Test Performance

When students find that they are deficient in certain competencies and these deficiencies hinder better performance, they may look for opportunities to improve on their weak areas. This trend is very prevalent in CSUF because its students are varied in their competencies. And so if students have the opportunity to access course materials from a supplementary web site seven days a week, 24 hours a day, and are also able to communicate with fellow students independent of time and distance, they could expect to do better in tests and generally in the entire course by using this opportunity. Haworth (1998, 1999) found this to be true in his studies. In another study, Slattery (1999) found that there were more "A" grades in his class when using a web site. Hester (1999) found that interactive sample tests on the web site were effective in that they led to improved student performance in exams. In comparing average test scores in courses that were taught entirely face-to-face with courses that were taught entirely online, students performed equally well in online format (King and Hildreath 2000) or better in online courses than their in-class colleagues (Arbaugh 2000, Bartlett, 1997; Heines and Hulse, 1996; Kabat and Friedel, 1990; Raymond III, 2001). Therefore, if a course is supplemented with a web site, students are likely to use the site more often in the hope of doing better.

Purpose of this Study

A review of the literature in the area of distance education shows that there are plenty of studies that have examined the overall reactions to and satisfaction with courses that were offered entirely online. Some studies have made an attempt to compare face-to-face with online instruction and have discussed their relative merits and demerits. However, there is a paucity of empirical studies to examine the specific motivations of students for visiting course dedicated web sites. Therefore, it is worth studying if students are visiting a course dedicated web site:

1. To obtain more resources from the instructor.
2. To communicate and collaborate with fellow students who may be hard to reach.
3. To improve their test performance.
4. To catch up on lost ground because of having to be absent from class.

Methods

The purpose of this section is to describe the research site, the methods used to collect and analyze data and the variables mentioned in the questionnaire. This research was conducted at the California State University, Fullerton, California. Subjects in this study consisted of undergraduate students in the Information Systems curriculum in the Information and Decision Sciences Department. The sample consisted of students from two sections each of Systems Analysis and Design and Database Design courses.

A paper and pencil questionnaire was distributed to every student in the class on the day of the final exam and was collected right away. Every student volunteered to answer the questionnaire. This gave a response rate of 100%. In all, 103 students responded to the survey. Data thus gathered was analyzed in various ways to answer the research questions posited.

The questionnaire employed in this study was exclusively designed to capture the data relevant to this study. All the questions were newly generated and the questionnaire was completed after several iterations of refinement.

Section One of the questionnaire contained items to determine the biographic profile of the student such as: year in school, major, ethnicity, age, gender, work experience, computer ownership and access to the Internet. Following these questions, statements describing the conditions of course environment were stated and students were asked to express the extent of their agreement to these conditions. Responses to these statements were captured on a Likert scale ranging from (1) Strongly Disagree, (2) Disagree, (3) Agree, and (4) Strongly Agree. Usefulness of the web site was measured by the following variables: (1) "web site usefulness in this class," (2) "useful in downloading notes," (3) "useful in setting up an account for my project," (4) "helped to improve academic performance"was used to measure the need to improve test performance, and (5) "instructor's notes helped." User friendliness of the web site was assessed by variable number six: "I found the web site to be user-friendly." The need to collaborate and be independent of time was measured by variable number seven: "Useful for exchanging notes with people whose working hours are different from mine." Variable number eight: "No need to attend class, since most of the things I need to know are available on the web site" measured the issue of attendance.

The courses considered in this study were taught by an instructor who had been successful in teaching these courses face-to-face for a few semesters prior to the semester in which this study was conducted. And so standards from the traditional classroom were available and considered to be suitable measures (Karuppan and Karuppan, 1999; Webb, 2001) for measurement of the Web assisted course.

Results

Fifty seven percent of the sample consisted of males and 42% were females. Eight percent of the students were Information Systems majors, 11% were Management Science majors. Eighty-two percent of the students were between the ages of 18 to 28, 15% were between 29 and 39 years, and 2% were between 40 to 49 years, reflecting the traditional college student population of a public institution. Ninety seven percent of the students said that they owned a computer and had access to the Internet either from home or school or both.

Table 1 summarizes the results of the regression analysis that was done to identify significant independent variables that could explain the variation in web site use. Twenty-nine percent of the variation in web site use was explained by eight independent variables. The most important predicator was "web site usefulness in this class," the second was "useful in downloading notes," the third was "useful in setting up an account for my project," the fourth was "helped to improve academic performance," the fifth was "instructor's notes helped," the sixth was "I found the web site to be user-friendly," the next was "useful for exchanging notes with people whose working hours are different from mine," and the last was "no need to attend class, since most of the things I need to know

Table 1. Reasons for Using the Web Site

Var num	Variable name
1	Web site usefulness
2	Useful in downloading notes
3	Useful in setting up an account for my project
4	Improve academic performance
5	Instructor's notes helped
6	Web site is user-friendly
7	Web site is useful in exchanging notes
8	No need to attend class

is available on the web site." It is worth noting that all the variables that were entered into the equation did prove to be significant predictors of web site use.

A two-tailed Pearson correlation between the web site use and the score shows that these variables were highly correlated among the "A" group of students ($r = .43$) followed by "C" group ($r = .38$) and then followed by "B" students ($r = .13$).

Table 2 shows the mean score among the different user groups where the sample was broken down by the degree of use, such as "use every day, several times per week, once a week, several times a month and never." The mean score was 84% among those who said that they never used the web site, followed by 81% among those who said that they used the web site at least once a week and those who said that they used it every day. The mean score was 79% and 76% among those who said that they used the site several times a month and several times per week, respectively.

Table 2. Mean Score Among Different Groups of Users

Web site users	Percent	Mean Score
Every day	2.4	81
Several times per week	16	76
Once a week	35	81
Several times a month	42	79
Never	6	84

N = 84

Discussion

This study treated the issues discussed in the information technology and distance education literatures to an empirical assessment. Results from this study lend support to the diffusion of innovation theory which argues that perceived usefulness of a technology and its user friendliness can lead to a decision to adopt that technology. Support for this comes from the variables "Useful in downloading instructor's notes," "Useful in setting up an account," "Web site is useful in this class," "Instructor's notes helped," and "This web site is user-friendly," all of which approached significance. Further, these results support prior research by Arbaugh (2000) who found that perceived usefulness of a course web site led to its use.

The need to use a web site to collaborate with fellow students by overcoming the constraints of time and distance found support in this study and affirmed earlier findings by Ragothanman and Hoadly (1999). The need to have an account for the group project and be able to exchange notes with others found support in this study and played an important role in determining the use of the web site. The variable "Using a web site in this class helped me improve my academic performance," was a significant predictor of web site use and ranked fourth in the order of importance. This affirmed prior research by Haworth (1998) and Slattery (1999), who found that students in courses that were offered entirely online performed very well. The variable "No need to attend class since everything I need to know is on the web site" was significant in predicting the web site use, since it entered the regression equation. A closer look at the number of people who held this view showed that 33% strongly disagreed, 45% disagreed, 15% agreed, and 5% strongly agreed. And so there were some students who stayed away from class and might have relied entirely on the notes on the web site, but they were in the minority. The majority of the students did not perceive the web site as an invitation to miss class. On the other hand, students downloaded the notes prior to the lectures and brought them to class and lab sessions and elaborated on them during classroom lectures, as was observed by the researcher. The attendance was as normal as it was in earlier semesters when the web site was not used. This confirms the findings in an earlier research (Karuppan and Karuppan, 1999).

The student mean use of the web site among the different grade categories were more or less the same. The positive correlation between the variables web site use and score among "A" ($r = .43$) and "C" ($r = .38$) students suggests that these students were more concerned in improving their grades, rather than willing to accept the status quo. It is possible that the "A" students were concerned about maintaining their superior performance and, therefore, used the opportunity of a web site to do better. In the case of the "C" students, their concern to strive to do better might have been their motivation to use the web site.

Observation of the mean scores (Table 2) from among the different user groups categorized from "used every day" to "never used" suggests that the highest scores were earned by those who said that they never used the web site. It is possible that these students relied entirely on class notes and studied from the text and managed to get an "A" grade. This observation reflects the power of empirical evaluation to be able to uncover naturally occurring phenomena.

The overall course grade point average was within the normal range of grade point averages in the courses that were taught face-to-face in prior semesters by the same instructor. There was no evidence of an unusually high grade point average on account of using the web site. But, in observing the pattern of visits to the web site among the different grade categories, "A" students used the web site more than the "B" students and so did the "C" students.

The study reveals that students look for web sites that are useful, friendly to access, able to provide the ability to communicate and able to provide an opportunity to catch up and do better on exams. Therefore, the following suggestions are made for each of these needs:

- As Weigel (2000) suggests, make the web site rich in content so that the students perceive the site to be useful. Other than making just the instructor's notes available, set up hyperlinks for students to reach the web sites of other professors and other sources of relevant information. For instance, setting up a link to the local chapter of the Oracle users group can help students of SQL programming language take part in discussions regarding database programming. Links to web sites of various vendors of drawing tools and CASE (Computer Assisted Software Engineering) tools can help students to download trial versions of the software and try them and acquire additional skills. Select textbooks which have excellent web site support for their material (Webb 2001; McEwen, 2001). Publishers like Course Technologies and Prentice Hall assist authors by providing additional notes, power point slides, chapter quizzes, tutorials, exercises and executable program code, to demonstrate a variety of programming topics and issues discussed in the text.

- Instructors can design web sites with specific student needs in mind. Challenging activities at different levels should be made available for students in different grade categories ("A" thru "F"). Interactive self-test and review modules, access to past exams and tutorials can collectively have an impact on student performance. In order to see significant differences in scores, instructors can set up a process whereby students accumulate points for completing certain exercises from the web site. This would force students to pursue the web site more actively and can lead to better performance and higher scores as was evidenced by Webb (2001). Having a web site which is close to face-to-face instruction and then expecting students to voluntarily visit the web site may not lead to significant differences in test scores or grade point averages, as was evidenced in this study.

Although this study showed that a supplementary web site does not deter classroom attendance completely, it would be a wise idea for instructors to counter balance classroom attendance with visits to the web site. Planning on class discussion about issues identified in the "hot site of the week" on the web site can make students feel the need to attend class. For instance, a plan to discuss in class the successful implementation of the new operating systems from Microsoft, such as Windows XP, can excite students to go to Microsoft's web site and read about any latest news about its successful implementation stories. Further, informing the students that some questions from the discussion will be on the test can motivate students to visit the web site, attend class and prepare for tests. The downside effect of such a suggestion is that planning on rich content and judiciously balancing lectures with web site material can place a huge demand on the instructor's time and instructors need to be aware of this and budget their time accordingly.

- Select course software that is user-friendly and which facilitates communication among individuals and groups, and allows threaded discussions. During lectures, pay attention to the communication patterns that emerge in the class and take a proactive stance in communicating (Hartman and colleagues, 2002) using the system's messaging tools. The reason for this being that there are always a few students in class who need clarification on the same issues many times. For example, students can e-mail about exam dates or assignment due dates even though this information can be found on the syllabus and on the web site. By placing timely announcements on your web site, you can decrease the volume of incoming messages and save yourself the "In-box shock," as Hartman and colleagues (2002) put it.

- Regarding selecting user-friendly software, university administrators also have some suggestions to take. Universities must take a leading role and must standardize the software to be used for developing web sites for all the courses in the university curriculum. The reason for this suggestion is that students would then have to be familiar with only one courseware. For instance, at CSUF the university had leases with both WebCT and BlackBoard for sometime. During that period, students had to be familiar with both WebCT and BlackBoard, because some instructors used WebCT while others used BlackBoard. This placed a big demand on students. Further, there were constant complaints from faculty and students about WebCT not being user friendly. As a result, the university decided to discontinue its lease with WebCT and opted to go with only BlackBoard. Now, with all of the Web-based courses available only on BlackBoard, students need to be acquainted with only one courseware and the entire collection of Web-based courses in the school share the same features.

Strength and Weaknesses of this Study

This study was an unobtrusive study of student perception of a course-dedicated web site. The natural setting greatly enhanced the basis of external validity. In studies trying to explain communication patters and behaviors, field settings are more appropriate than experimental designs with control groups. The fact that the factors chosen for examination are deeply anchored in theory, lends some protection to the problem of generalization (Kidder and Judd, 1986). The survey instrument captured perceptual data in which measurement errors in responses can place limitations on the study. The web site used in this study received verbal feedback while it was in the process of construction in prior semesters and so a formal pretest of the site was not undertaken. The course grade point average in the same courses taught in prior semesters in a face-to-face mode by the same instructor was used as a standard of measurement in trying to compare the overall performance in the Web assisted class with the face-to-face class. Since this study sought to examine, among other issues, the relative web site usage pattern between the different grade categories, i.e., whether the "A" students used the site more than the "B" students, and so on, the need for a control group was not seen as necessary. However, the absence of results from pre-testing the web site and absence of results from a control group can be seen as limitations from a strict statistical point of view.

Furthermore, results in this study are based on students at one specific university and may not be generalizable to students at other universities. Finally, due to small sample size, care must be taken in interpreting the results of this study. However, they do shed light on the questions posed in this study and support the assertions and are consistent with the findings of previous research. Results from this study can be insightful for the sister campuses in the California State University system.

Directions for Future Research

Future research activities can be directed towards improving the richness of educational software. Studies can attempt to match the richness of the electronic media with the learning task at hand. For instance, if the task is a case analysis that requires discussion and sharing various points of view, then electronic discussion boards would be suitable. How comfortable are students communicating through this media can be a worthy study. Do male and female students feel the same way about expressing themselves through electronic discussion boards? Are language skills influential in the use of this media or are they a deterrent? These are some questions to ask.

As the development of web sites snowball into well developed, effective and matured sites, an issue that assumes significance in this context is the right to ownership of effective web sites that members of the faculty might have developed painstakingly. Although faculty may think they own the materials on the Web, there is considerable debate as to who has the right to course materials that have been developed with university resources and release time. Many instructors try to overcome this issue by adopting textbooks that have online course materials. More and more publishers are now beginning to provide online courses for the hard copy textbooks that they market and many of these vendors have contracted with BlackBoard, which offers the overall structure of an academic web site and the mechanics of delivering it, leaving the course content decisions to the instructors. So, as this movement gains momentum, more research needs to be done as to how universities should handle the issue of intellectual property rights.

Conclusion

There is no naysaying the fact that online education offers higher education to those populations that are deprived of educational opportunities. As Rosenbaum (2001) puts it, online learning "democratizes education" by placing resources of prestigious universities with in the reach of those who would otherwise not have access to it. But, whether it should replace traditional face-to-face instruction at the undergraduate level at a public institution is debatable. However, introducing undergraduates to online learning would serve them very well because online learning serves as the back bone of continuing education and life long learning and it is very likely that these students will enter a phase in their lives when they have to renew their knowledge and skill. And so, to the question of "how can technology be incorporated in the learning paradigm?" The answer, according to this researcher, is to select a paradigm that blends the best in online learning and face-to-face learning.

Students in this study visited their course-assisted web site in order to find useful resources which could be accessed by user friendly software, to find opportunities to communicate electronically with their professor and classmates, to look for opportunities to improve their academic performance and to "catch up" on lost ground. Although, they did not all earn A's, they saw the web site as a valuable complement to their face-to-face lectures. Results in this study were used to suggest design strategies that could assist in building richer web sites which could eventually lead to a richer learning environment.

References

Angulo, A. J. and Bruce, M. (1999, Winter). Student perceptions of supplemental Web-based instruction. *Innovative Higher Education*, 24(2), 105-25.

Arbaugh, J. B. (2000, February). Virtual classroom characteristics and student satisfaction with Internet-based MBA courses. *Journal of Management Education*, 24(1), 32-54.

Atkinson, M. and Kydd, C. (1997). Individual characteristics associated with World Wide Web use: An empirical study of playfulness and motivation. *The DATA BASE for advance in Information Systems*, 28(2), 53-62.

Bagozzi, R. P., Davis, F. D. and Warshaw, P. R. (1992). Development and test of a theory of technological learning and usage. *Human Relations*, 45(7), 659-686.

Bailey, R. A., Chow, W. C. and Haddad, M. K. (1999, January/February). Continuous improvement in business education: Insights from the for-profit sector and business school deans. *Journal of Education for Business*, 75, 165-180.

Bartlett, T. (1997, Oct.). The hottest campus on the Internet. *Business Week*, 3549, 77-80.

Brandon, D. P. and Hollingshead, A. B. (1999, April). Collaborative learning and computer-supported groups. *Communication Education*, 48(2), 109-26.

The Campus Computing Project (1999). *The Continuing Challenge of Instructional Integration and User Support*. Encino, CA.

DeSanctis, G. and Sheppard, B. (1999). Bridging distance, time and culture in executive MBA education. *Journal of Education for Business*, 157-160.

Eastman, J. K. and Swift, C. O. (2001). New horizons in distance education: Online learner centered marketing class. *Journal of Marketing Education*, 23(1), 25-35.

Evans, C. and Fan, T. P. (2002). Life long learning through the virtual university. *Campus Wide Information Systems*, 19(4), 127-134.

Evans, J. R. and Hasse, I. M. (2001). Online Business Education in the 21st century: An analysis of potential target markets. *Internet Research*, 11(3), 246-260.

Everett, R. (1983). *Diffusion of Innovations (3rd Ed)*. New York: Free Press.

Everett, R. (1986). *Communication Technology: The New Media in Society*. New York: Free Press.

Gefen, D. and Straub, D. W. (1997). Gender differences in the perception and use of e-mail. An extension to the technology acceptance model. *MIS Quarterly*, 21(4), 389-400.

Hartman, J., Lewis, J. S. and Powell, K. S. (2002). Inbox Shock: A study of electronic message volume in distance managerial communication course. *Business Communication Quarterly*, 65(3), 9-28.

Haworth, B. (1998). *Do students benefit from course based web sites?* Meeting of Economic Association in Lexington, Kentucky.

Haworth, B. (1999, September/October). An analysis of the determinants of students e-mail use. *Journal of Education for Business*, 75(1), 55-59.

Heines, R. A. and Hulse, D. B. (1996). Two way interactive television: An emerging technology for university level business school instruction. *Journal of Education for Business*, 71(2), 74-76.

Hester, J. B. (1999, Spring). Using a Web-based interactive test as a learning tool. *Journalism and Mass Communication Education*, 54(1), 35-41.

Kabat, E. J. and Friedel, J. (1990). *The development, pilot testing, and dissemination of a comprehensive evaluation model for assessing the effectiveness of a two-way interactive distance learning systems.* (ERIC Document Reproduction Service No. Ed 332 690).

Karuppan, C. and Karuppan, M. (1999, September). Empirically based guidelines for developing teaching materials on the Web. *Business Communication Quarterly*, 62(3), 37-45.

Kemelgor, B. H., Johnson, S. D. and Srinivasan, S. (2000, January/February). Forces driving organizational change: A business school perspective. *Journal of Education for Business*, 75(3), 133-37.

Kidder, L. H. and Judd, C. M. (1986). *Research Methods in Social Relations (5th Edition)*. Holt, Rinehart and Winston.

King, P. and Hildreth, D. (2000). Internet courses are they worth the effort? *Journal of Science Teaching*, XXXI(2), 112-115.

Lahey, L. (2002). Virtual school drop-ins. *Computing Canada*, 28(20), 26-29.

Major online effort at MIT. (2001). *Campus Wide Information Systems*, 18(3), 99-100.

Mangan, K. S. (1999, January). Top business schools seek to ride a bull market in online MBAs. *Chronicle of Higher Education,* 45(19), A27-28.

McEwen, B. C. (2001). Web assisted and online learning. *Business Communication Quarterly*, 64(2), 98-103.

NEA (National Education Association) (2000, June). *A Survey of Traditional and Distance Learning Higher Education Members. Ed at a Distance.* Retrieved from the World Wide Web at: WWW.nea.org.

Okula, S. (1999, February). Going the distance: A new avenue for learning. *Business Education Forum*, 53(3), 7-10.

Ponzurick, T. G., France, K. R and Logar, C. M. (2000). Delivering graduate marketing education: An analysis of face to face versus distance education. *Marketing Education*, 22(3), 180-188.

Rad, P. F. (2002). Distance education. *Cost Engineering*, 44(6), 9-11.

Ragothaman, S. and Hoadley, D. (1997, March/April). Integrating the Internet and the World Wide Web into the business classroom: A synthesis. *The Journal of Education for Business,* 72(4), 213-216.

Raymond III, F. B. (2000). Delivering distance education through technology: A pioneer's experience. *Campus-Wide Information Systems*, 17(2), 49-55.

Reisman, S., Dear, R. and Edge, D. (2001). Evolution of Web-based distance learning strategies. *The International Journal of Educational Management,* 15(5), 245-251.

Rosenbaum, D. B. (2001, May). E-learning beckons busy professionals. *Engineering News Review*, 246(21), 38-42.

Schwartz, H. (1995). Computers and urban commuters in an introductory literature class. In M. Collins and Z. L. Berge (Eds.), *Computer Mediated Communication and the Online Classroom* (Vol. 2: Higher Education).

Slattery, M. J. (1998). Developing a Web-assisted class: An interview with Mark Mitchell. *Teaching of Psychology*, 25(2), 152-155.

Surrency, D. P., Bishop, D. N. and Hendrix, E. H. (2001, Winter). The distance learning composition classroom: Pedagogical and administrative concerns. *ADE Bulletin*, 127, 55-59.

Webb, J. P. (2001). Technology: A tool for the learning environment. *Campus Wide Information Systems,* 18(2), 73-78.

Witmer, D. F. (1998). Introduction to computer-mediated communication: A master syllabus for teaching communication technology. *Communication Education,* 47, 162-173.

World Bank: Major force in distance education. (2001). *Campus Wide Information Systems*, 18(2), 52-54.

About the Authors

Caroline Howard has an M.B.A. from The Wharton School, University of Pennsylvania, and a Ph.D. in Management Information Systems from the University of California, Irvine. She is an associate professor at Touro University and president of Techknowledge-E Systems (USA). Prior to Emory, she was on the IT faculty at the University of Colorado, Colorado Springs. In addition to being an editor of *The Design and Management of Effective Distance Learning Programs* (Idea Group Publishing, January 2002), Caroline recently published *Winning the Net Game: Becoming Profitable Now that the Web Rules Have Changed* (Entrepreneur Press, June 2002). She has written numerous distance education and IT articles.

Karen Schenk holds a Ph.D. in Information Technology and an M.B.A. in Finance from the University of California, Irvine. She has been a professor of Information Systems at the University of Redlands, California, and at North Carolina State University. She has taught numerous courses in information technology, decision support systems, and systems design. Her publications have focused on distance education, lifelong learning, and human-computer interfaces. She has co-edited the book, *The Design and Management of Effective Distance Learning Programs* (Idea Group Publishing, January 2002). She is currently senior partner of K. D. Schenk and Associates Consulting (USA), working with companies and clients on IT issues and distance learning.

Richard Discenza is a professor of Production Management and Information Systems in the College of Business and Administration, University of Colorado at Colorado Springs (USA). He received his B.S.F. in Forestry from Northern Arizona University, an M.B.A. from Syracuse University, and a Ph.D. in Management from University of Oklahoma. Dr. Discenza was formerly dean of the college, where he helped establish and oversaw the development of the distance M.B.A. program. His current research focuses on business process reengineering, distance education, project management, and supply chain management. He has published numerous articles in professional and academic journals and is a member of APICS, the Academy of Management, and PMI. This is Dick's second publication as an editor for Idea Group Publishing. In 2002, he edited *The Design and Management of Effective Distance Learning Programs* (IGP, January 2002).

*   *   *

Charles E. Beck is an associate professor of Management and Communication at the University of Colorado at Colorado Springs, USA. As an Air Force officer, his career began in aircraft maintenance, but included teaching at the Air Force and Naval Academies, and the AF Institute of Technology. He began with the University of Colorado as director of the MS in Technical Communication prior to moving to business. Chuck has published in numerous journals and proceedings, and Prentice-Hall published his book *Managerial Communication: Bridging Theory and Practice*. He has served as a consultant to businesses in Dayton, Washington D.C., and in Colorado.

Judith V. Boettcher is a consultant and author in technology, instruction and online learning. She has consulted with a wide range of major universities and organizations on projects such as Designing for Learning (USA), faculty development, and future trends of uses of technology. In addition to consulting, Judith is currently chairing the Syllabus 2003 conference and serving on the academic advisory committee of ICUS, an e-learning company in Singapore. She can be reached at judith@designingforlearning.info.

Elizabeth A. Buchanan is assistant professor and co-director of the Center for Information Policy Research at the School of Information Studies, University of Wisconsin-Milwaukee, USA. She researches, teaches, and writes in the areas of distance education and online learning, information ethics, information policy, and research ethics.

Gregory Claeys is a Metis/Cree from central Canada and is an education consultant for universities, colleges and first nation communities in British Columbia. As a former student of the University of Victoria's Administration of Aboriginal Governments Program in the School of Public Administration, Greg's interests include bridging the gap between mainstream society and aboriginal people by offering educational opportunities that bring different world views, visions, values, beliefs and perspectives into the framework of educational disciplines and creating opportunities for both aboriginal and non-aboriginal people.

Kim E. Dooley is an associate professor in the Department of Agricultural Education at Texas A&M University, USA. She has conducted numerous professional presentations and training programs around the globe. Her publications include a chapter in *Distance Training*, which received the Wedemeyer Award from the Association of Continuing Education. She has served on many university/system committees and advisory boards, including the American Distance Education Consortium International Taskforce. She was also the 1999 recipient of the Montague Teaching Scholar Award and the 2002 International Excellence Award at Texas A&M University. She is an active member of the American Association of Agriculture Education (AAAE) and the Association for International Agricultural and Extension Education. She received the Outstanding Young Member Award for the Southern Region of AAAE in 2002 and has won the Outstanding Paper Presentation three times (National, 2000, Western Region, 2000, and Southern Region, 2003). She also serves on the editorial board for the *Journal of Agricultural Education*.

Jonathan Frank is an associate professor of Information Systems and Operations Management, Sawyer School of Management, Suffolk University, Boston, USA. He received his Ph.D. from the University of Strathclyde in Glasgow, UK, in 1978. He has taught courses and led workshops on e-commerce development and Web design at universities in Europe, Canada, the US, Africa and the South Pacific. His current research focuses on the management of cross-cultural virtual teams. He has published in the management information systems and distance education fields.

Svenja Hagenhoff, Ph.D., born July 11,1971 in Muenster, Germany, worked after getting her German high school degree in the Electrochemical Industry as a commercial clerk from 1991 to 1992. From 1992-1997 she studied Business Economics at the Georg-August-University Goettingen, Germany. Along the way she did some practical trainings at Deutsche Welle TV in the editorial offices and at Hoechst (today, Aventis) in controlling and computer engineering. From 1997-2002 she worked as a research assistant at the chair of Information Management, Dept. 2 at the Georg-August-University Göttingen, Germany. She did projects at the areas of e-learning and information technology in financial services. In addition, she holds a teaching position at the Leibniz-Akademie Hannover, Germany, where she taught information management from 1997-2001. In 2001 she received her Ph.D. (Dr. rer. pol.). Since September 2002, she has served as scientific assistant and leader of two research teams at the chair of Information Management at the University of Goettingen in order to get the *venia legendi* (Habilitation). Her research interests are innovation management and knowledge and education management.

Oliver Kamin, Dipl.-Hdl., born May 12, 1971, in Gronau, Germany, received his German High School degree and his professional training from 1991-1993 as an industrial clerk at Deutsche Tiefbohr-AG in Bad Bentheim, Germany. Afterwards he did his military-service. From 1994-1997, he studied business economics at the Justus-Liebig-University, Gießen, Germany. After his diploma preliminary test, he changed to the Georg-August University, Göttingen, Germany, to continue his academic formation in econonomics and business education. After getting his diploma, he worked at the Institute for Economics and Business Education at the Georg-August-University, Göttingen, Germany. Since 2001, he has worked as a research assistant at the Institute of Information Systems, Dept. II, at the Georg-August-University, Göttingen, Germany.

Michaela Knust, Dipl.-Hdl., born February 15, 1972, in Lower Saxony, Germany, completed her vocational training as industrial clerk in 1994 at H. Butting GmbH & Co. KG, Wittingen, Germany. Afterwards, she was working for two years as commercial clerk at the Export Sales of H. Butting GmbH & Co. KG. From 1996-2001 she studied Business Education at the Georg-August-University in Göttingen, Germany. In 1998, she did her three-month-practical training at United Pipelines Ltd., in Warrington, UK, during the semester break. Since 2002 she has worked as a member of the education network WINFO*Line* and is completing her doctorate in economics at the chair of Information Management, Dept. 2 at the Georg-August University, Göttingen, Germany.

Malini Krishnamurthi, Ph.D., is a lecturer in Information Systems and Decision Sciences at the California State University, Fullerton, California, USA, where she teaches undergraduate and graduate courses in database design and programming with Oracle, systems analysis and design, data warehousing and knowledge management. Her research interests include organizational impacts of IT, strategic value of IT and the IS curriculum. She has been a reviewer and contributor to several conference proceedings and professional journals.

Yair Levy is instructor of Information Systems and Director of Online Learning at the College of Business Administration at Florida International University, USA. Prior to joining the college, he assisted NASA in developing e-learning platforms and management of Internet and Web infrastructures. Yair is a Ph.D. candidate in MIS at FIU and is finishing his dissertation in the area of online learning systems. He earned his bachelor's degree in Aerospace Engineering from the Technion (Israel Institute of Technology). He has received his M.B.A. with an MIS concentration from FIU. His current research interests include value of information systems, value of online learning systems, IS and online learning effectiveness. He served as a referee research reviewer for several international scientific journals and conferences (AMC-SIGMIS, ICIS, HICSS, AMCIS, and ICPAKM). Yair's teaching interests and courses taught include: telecommunications and networking, Web management, and e-commerce technologies for managers in the Master of Science in MIS program and for undergraduate MIS majors.

James R. Lindner is an assistant professor in the Department of Agricultural Education at Texas A&M University, USA. He has authored or co-authored more than 100 refereed research papers and two textbooks. His research focuses on planning and needs assessment, management of human resources, and distance education. His article "Understanding Employee Motivation" was named the most frequently viewed *Journal of Extension* article. Within the department, Dr. Lindner is helping to develop and deliver the Masters of Agriculture at a distance program and the Doc @ Distance program. He has received numerous honors and awards for presentations of research findings at international and national conferences and was recently named the Outstanding Young Agricultural Educator by the American Association for Agricultural Education.

Wm. Benjamin Martz, Jr., is an associate professor of Information Systems at the University of Colorado at Colorado Springs (USA). Ben's teaching interests include e-business, software development, groupware and team-based problem solving. He received his B.B.A. in Marketing from the College of William and Mary and his M.S. in Management Information Systems (MIS) and his Ph.D. in Business, with an emphasis in MIS, from the University of Arizona. Ben was one of the founding members, as well as president and COO, of Ventana Corporation — a technology, spin-off firm from the University of Arizona — incorporated to commercialize the groupware software product

GroupSystems. In 1994, GroupSystems won *PC Magazine's* Editor's Choice award for best Electronic Meeting System software. Ben has published his groupware research in *MIS Quarterly*, *Decision Support Systems,* and the *Journal of Management Information Systems* and his student learning environment research in the *Decision Sciences Journal of Innovative Education*, *Journal of Cooperative Education* and *Journal of Computer Information Systems*.

Michelle M. Ramim is an MIS consultant helping corporations and educational institutions on varieties of Information System implementations including new e-commerce and online learning platforms. Michelle previously served as the Director of the Instructional Technology Center at Florida Memorial College in Miami, Florida. She has extensive experience in consulting. Michelle directed the development and implementations of several Information Systems, including promotional and interactive web sites, and online learning web sites for several educational institutions utilizing WebCT, and other online learning platforms. She is currently admitted to a Ph.D. program in Information Systems at Nova Southeastern University. Michelle is a frequent speaker at national and international meetings on management information systems and online learning topics. She earned her bachelor's degree from Barry University in Miami, Florida. Michelle has received her M.B.A. from Florida International University.

Venkateshwar K. Reddy is the associate dean and an associate professor of Finance at the University of Colorado at Colorado Springs, USA. Dr. Reddy received his master's and doctorate degrees from the Pennsylvania State University. Dr. Reddy's interest and efforts led to building a state of the art, nationally ranked Distance M.B.A. program at the University of Colorado at Colorado Springs. The program averages more than 1,000 enrollments per year up from less than 500 when Dr. Reddy took over the program. Dr. Reddy also teaches in the CU's Executive M.B.A. program, which attracts mid- to high-level managers, entrepreneurs, and doctors among other professions. Twice, he has received the Outstanding Teacher Award from the College of Business for his teaching efforts. Professor Reddy's primary research and teaching interests are in the areas of investments and corporate finance. Dr. Reddy's research work has appeared in several finance and economics journals. His work on mutual fund investing strategies, published in the *Journal of Financial Planning* this year, has been quoted in the *New York Times* and other newspapers and major magazines around the country.

Lance J. Richards joined the Department of Petroleum Engineering at Texas A&M University (USA) in August 2002 as a distance learning coordinator. He received a B.S. degree in Agricultural Education from Texas A&M University in December 2001 and will complete an M.S. in Agricultural Education at Texas A&M University in December 2003. He is also certified to teach in the state of Texas. Lance's background is in teaching, educational theory, instructional design, the management of instructional telecommunication systems, planning and needs assessment, and in the delivery of instruction at a distance. His research includes comparisons of distance delivery methods, the application of new technology to distance programs, course design principles for distance education, needs assessment for programs seeking to offer distance education, and the evaluation of existing distance degree programs.

Evan T. Robinson, R.Ph., Ph.D., director, Division of Technology in Education, associate professor, Department of Biopharmaceutical Sciences, Shenandoah University School of Pharmacy (USA). Evan received his B.S. degree in Pharmacy and M.S. in Pharmacy Administration from St. Louis College of Pharmacy. He received his Ph.D. in Pharmacy Administration from Auburn University, the Department of Pharmacy Care Systems. Since joining Shenandoah University his responsibilities have included: overseeing an Internet-based, non-traditional doctor of pharmacy program started in September 1998 with a steady-state enrollment of 120 students and 175 graduates; evaluating onsite and offsite applications of technology for teaching and learning; the development and administration of certificate and continuing pharmaceutical education programs and teaching within the school of pharmacy.

Richard Ryan is an associate professor at the University of Oklahoma in the College of Architecture Construction Science Division (USA). He has been actively involved in teaching and using information technology for construction applications since 1992. He has done Web development and technology consulting for construction companies. In Spring 1998, he offered the first complete semester length online construction class (cns4913online, Construction Equipment and Methods) to other construction programs. His experiences have fostered many observations pertaining to creation, organization, promotion and administration of web-based distance learning. Currently, he is working with continuing education at the University of Oklahoma to host a construction administration online portal to be purchased by students. The portal will offer a study guide, links and related activities.

Gilly Salmon is an academic member of the Centre for Innovation, Knowledge and Enterprise at the Open University Business School and Visiting Professor at Glasgow Caledonian Business School, both in the UK. She chairs the OUBS's large online Professional Certificate in Management. She has been involved in

online teaching and learning since the 1980s. She has two research degrees —
one in Change Management and one in E-training (she says she needs both in the
e-world!). Her recent books are called *E-tivities* and *E-moderating*. Web sites:
http://www.atimod.com/presentations, http://www.atimod.com/e-moderating,
http://www.e-tivities.com.

Karen Sangermano is the program manager for the University of Colorado
(USA) - Colorado Springs (UCCS) Distance MBA program. She has worked
with the UCCS Distance M.B.A. from its inception and coordinates many of the
day-to-day functions of the program, including discussions with the students
about their impressions of the course and the overall program. Karen holds a
B.S.B.A. in Accounting from CU-Colorado Springs.

Gary R. Schornack is a faculty member and director of mentorship programs
at the University of Colorado's Denver Business School (USA). Gary is a
frequent speaker and author of more than 40 papers in the past three years. His
national and international presentations have included a variety of topics:
marketing competition, communications strategies, distance education, mentorship,
and knowledge management. As a marketing and management consultant, he
has worked with many companies including: Coors Brewing Co., Qwest, and the
U.S. Office of Education. His web site received the Golden Web Award four
years in a row for one of the best in the nation.

Morgan M. Shepherd is an associate professor of Information Systems at the
University of Colorado at Colorado Springs, USA. Morgan spent 10 years in
industry, most of that time with IBM. His last position with IBM was as a
technical network designer. He earned his Ph.D. from the University of Arizona
in 1995 and has been teaching for the I/S department at the University of
Colorado in Colorado Springs since then. His primary teaching emphasis is in
telecommunications at the graduate and undergraduate level. He has also taught
numerous courses on computer literacy, web design and systems analysis and
design. In addition, he has been teaching courses via distance education for
several years. His primary research emphasis is on making distributed groups
productive and applying this research to business as well as education. His
research has appeared in the *Journal of Management Information Systems*,
Informatica, and *Journal of Computer Information Systems*.

Janet Toland is a lecturer in Information Systems at Victoria University of
Wellington, New Zealand. She has 20 years experience in the field of IS, both
in industry, and as an academic. She has worked in the UK, Botswana, Fiji and

New Zealand. Her areas of research are systems analysis and design, virtual organizations, the virtual university, computer mediated communication, the digital divide and computer supported co-operative work. She is currently investigating the opportunities for e-commerce in the South Pacific, and the development of learning regions in New Zealand.

Murray Turoff is a distinguished professor of Information Systems at the New Jersey Institute of Technology, USA, and holder of the research chair for the Hurlburt Professor of Management Information Systems. Dr. Turoff has been involved in research, development, and evaluation of computer mediated communication systems since the late '60s. Since the early '80s he has been utilizing group-oriented communication systems to augment college courses and for offering distance versions of college level courses. He is a noted advocate for the use of modern group communications technology and appropriate learning methodologies for improving the quality of education at all levels.

Elizabeth Wellburn works as an instructional designer at Royal Roads University in Victoria, British Columbia, Canada. Her background includes projects involving design, development, implementation and research related to technology for education and training. She has worked in corporate, government and academic contexts and is an experienced facilitator of distributed learning courses. Elizabeth's enthusiasm for technology as an educational tool is based on a vision of sustainability and capacity building through enhanced collaboration. She believes that this is possible both within and across communities of learning through consultation and the active participation of community members.

Index

A

academic courses 5
action based e-moderator training 70
adult learning principles 107
advanced methods 100
anonymity 8
anti-globalization concerns 81
assessment strategies 110
assimilation 89
asynchronous group communication
    technologies 2
audit analysis 259
automated construction teaching 240
automated format 236

B

Blackboard 102
budget 285
business-to-business (B2B) models
    256
business-to-consumer (B2C) transac-
    tions 256

C

Central Queensland University (CQU)
    218
certificate design 23
classroom interaction 153
co-operational partners 303
collectivist cultures 215
college curriculum 322
commercialization 146
communication protocol 8
community-based distributed learning
    79
competency-based course 100
computer-based training (CBT) 191
computer-mediated communication 215
computer-mediated system 2
congruency 21
content facilitator 58
content-specific modules 197
copyright costs 256
corporate education 264
cost-oriented 302
course calendar 103

course content 103
course design 23
course management tool 102
course orientation 103
course planning 100
course syllabus 103
cultural background 215
cultural differences 213
cultural diversity 214
curriculum design 23

D

degree design 23
development team 288
didactical guidelines 191
distance education 119, 144, 178
distance education administration 154
distance education classes 1
distance education materials 253
distance learning 2, 213
distance learning strategies 324
distributed learning opportunities 79

E

e-learning 120, 190
e-learning materials 190
e-mail 213
e-moderating 55
e-moderator 56
e-moderator training 62
e-store 237
education industry 145
education networks 302
education service providers 303
educational consumerism 13
educational environment changes 147
educational integration 127
educational process model 120, 122
educational technology 213
efficiency of teaching 1
electronic transcripts 2

F

face-to face learning 164
face-to-face classroom 1

face-to-face courses 1, 323
faculty 6
faculty development costs 255
faculty instruction costs 256
financial incentive 237
funding 285
funding crises 282
futures project 83

G

game approach 238
globalization and educational change 82
globalization and social change 82
globalized world 79
glocalization 86
group work 170

H

higher education 55, 163, 303

I

implementation approach 199
Indigenous communities 80
Indigenous pilot projects 87
individualist cultures 215
information and communication technologies (ICT) 281
information exchange 66
information management 305
information overload 9
information technologies 79
infrastructure design 23
input synergies 304
institutional design 23
institutional support 285
instructional delivery 322
instructional design 100, 236
instructor-based system 237
inter-university education network 302
interactive teaching 105
internal audit 258
Internet 236
Internet testing 177
Internet-based learning 302

J

journals 168

K

knowledge construction 67
knowledge transfer 219

L

learner-centered 100
learning curve effects 312
learning platform supplier 310
learning processes 56
learning support 113

M

market analysis 309
market stratification 260
market-oriented 302
membership status lists 8
modular Web-based teaching 190
motivation 64

N

National Education Association (NEA)
      325
new money 237

O

online assessment strategies 166
online conference 8
online course 57
online course design principles 99
online development 62
online educational content 253
online environments 165
online instruction 236
online learning 21, 100, 278
online meeting 172
online orientation 112
online socialisation 65
online students 163, 182
online teaching 56
online text-based discussion· 56

organizational education networks 306
output synergies 304

P

participation 167
pedagogy 132
peer-assessment 170
performance-based learning 236
phase-oriented approach 199
pilot program 290
portfolios 168
process facilitator 58
process optimizing 312
process synergies 304
production costs 255
program design 23

Q

quality of instruction 1
quasi-experimental studies 3

R

recurrent costs 255
return on investment 253
role control 8

S

scaffolding 62
scaling methods 9
self-assessment 169
self-directed learning 108
self-funding online learning program 279
site map 103
smart classroom 120
social construction model 220
social informatics 179
social pressures 4
South Pacific 213
stimulating learning 238
student assessment design 23
student-centered learning 108
student-centered system 237
support costs 256
synergy effects 304
system-specific modules 197

T

technological infrastructure 146
technologist 58
technology costs 255
technology training 112
technology-enhanced global education
      84
teleteaching 302
test performance 328
transformative income generation 257

U

UK Open University 57
unit and learning activity design 23
University of the South Pacific (USP)
      218
user-friendliness 326

V

virtual classroom 119
virtual universities 302
voting 9
Vygotskian theory of cognition 21

W

Web course tools 102
Web-based coursework 164
Web-based instruction 323
Web-based trainings (WBT) 191, 303
WebCT 102
WINFOLine 202, 303

NEW Titles
from Information Science Publishing

- **Instructional Design in the Real World: A View from the Trenches**
  Anne-Marie Armstrong
  ISBN: 1-59140-150-X; eISBN 1-59140-151-8, © 2004
- **Personal Web Usage in the Workplace: A Guide to Effective Human Resources Management**
  Murugan Anandarajan & Claire Simmers
  ISBN: 1-59140-148-8; eISBN 1-59140-149-6, © 2004
- **Social, Ethical and Policy Implications of Information Technology**
  Linda L. Brennan & Victoria Johnson
  ISBN: 1-59140-168-2; eISBN 1-59140-169-0, © 2004
- **Readings in Virtual Research Ethics: Issues and Controversies**
  Elizabeth A. Buchanan
  ISBN: 1-59140-152-6; eISBN 1-59140-153-4, © 2004
- **E-ffective Writing for e-Learning Environments**
  Katy Campbell
  ISBN: 1-59140-124-0; eISBN 1-59140-125-9, © 2004
- **Development and Management of Virtual Schools: Issues and Trends**
  Catherine Cavanaugh
  ISBN: 1-59140-154-2; eISBN 1-59140-155-0, © 2004
- **The Distance Education Evolution: Issues and Case Studies**
  Dominique Monolescu, Catherine Schifter & Linda Greenwood
  ISBN: 1-59140-120-8; eISBN 1-59140-121-6, © 2004
- **Distance Learning and University Effectiveness: Changing Educational Paradigms for Online Learning**
  Caroline Howard, Karen Schenk & Richard Discenza
  ISBN: 1-59140-178-X; eISBN 1-59140-179-8, © 2004
- **Managing Psychological Factors in Information Systems Work: An Orientation to Emotional Intelligence**
  Eugene Kaluzniacky
  ISBN: 1-59140-198-4; eISBN 1-59140-199-2, © 2004
- **Developing an Online Curriculum: Technologies and Techniques**
  Lynnette R. Porter
  ISBN: 1-59140-136-4; eISBN 1-59140-137-2, © 2004
- **Online Collaborative Learning: Theory and Practice**
  Tim S. Roberts
  ISBN: 1-59140-174-7; eISBN 1-59140-175-5, © 2004

Excellent additions to your institution's library! Recommend these titles to your librarian!

**To receive a copy of the Idea Group Inc. catalog, please contact 1/717-533-8845, fax 1/717-533-8661,or visit the IGI Online Bookstore at:
http://www.idea-group.com!**

Note: All IGI books are also available as ebooks on netlibrary.com as well as other ebook sources. Contact Ms. Carrie Skovrinskie at <cskovrinskie@idea-group.com> to receive a complete list of sources where you can obtain ebook information or IGP titles.

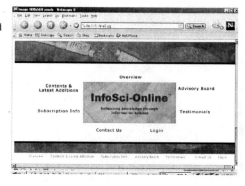